PhysioEx™ 4.0

Laboratory Simulations in Physiology

VERSION 4.0

TIMOTHY STABLER
Indiana University Northwest

PETER ZAO
North Idaho College

MARCIA C. GIBSON
University of Wisconsin–Madison
(Histology Review Supplement)

San Francisco Boston New York
Cape Town Hong Kong London Madrid Mexico City
Montreal Munich Paris Singapore Sidney Tokyo Toronto

Publisher: Daryl Fox
Project Editor: Barbara Yien
Managing Editor: Wendy Earl
Production Supervisor: Janet Vail

Benjamin Cummings gratefully acknowledges Carolina Biological Supply for the use of numerous histology images found on the PhysioEx CD-ROM.

PhysioEx™ 4.0 Products

For Anatomy & Physiology

PhysioEx™ 4.0
Laboratory Simulations in Physiology (CD Version):
order ISBN 0-8053-6144-8

PhysioEx™ 4.0
Laboratory Simulations in Physiology (Web Version):
order ISBN 0-8053-6150-2

Instructor's Guide for PhysioEx™ 4.0:
order ISBN 0-8053-6155-3

For Human Physiology

PhysioEx™ for Human Physiology 2nd ed. (CD Version):
order ISBN 0-8053-6153-7

PhysioEx™ for Human Physiology 2nd ed. (Web Version):
order ISBN 0-8053-6151-0

Instructor's Guide for PhysioEx™ for Human Physiology 2nd ed.: order ISBN 0-8053-6154-5

To locate the Benjamin Cummings sales representative nearest you, visit **http://www.aw.com/replocator**

ISBN 0-8053-6146-4

4 5 6 7 8 9 10--MAL--06 05 04 03

www.aw.com/bc

CONTENTS

	Preface	P-2
Exercise 1	The Cell--Transport Mechanisms and Permeability	P-4
Exercise 2	Skeletal Muscle Physiology	P-15
Exercise 3	Neurophysiology of Nerve Impulses	P-26
Exercise 4	Endocrine System Physiology	P-36
Exercise 5	Cardiovascular Dynamics	P-47
Exercise 6	Frog Cardiovascular Physiology	P-58
Exercise 7	Respiratory System Mechanics	P-65
Exercise 8	Chemical and Physical Processes of Digestion	P-73
Exercise 9	Renal Physiology	P-84
Exercise 10	Acid-Base Balance	P-92
	Using the Histology Tutorial	P-103
	Review Sheet for Exercise 1	P-105
	Review Sheet for Exercise 2	P-109
	Review Sheet for Exercise 3	P-113
	Review Sheet for Exercise 4	P-117
	Review Sheet for Exercise 5	P-121
	Review Sheet for Exercise 6	P-125
	Review Sheet for Exercise 7	P-127
	Review Sheet for Exercise 8	P-131
	Review Sheet for Exercise 9	P-135
	Review Sheet for Exercise 10	P-139
	Histology Review Supplement	P-141

PREFACE

PhysioEx Version 4.0 consists of ten physiology lab simulations that may be used to supplement or replace wet labs. This easy-to-use software allows you to repeat labs as often as you like, perform experiments without harming live animals, and conduct experiments that may be difficult to perform in a wet lab environment due to time, cost, or safety concerns. You also have the flexibility to change the parameters of an experiment and observe how outcomes are affected. In addition, PhysioEx includes an extensive histology tutorial that allows you to study histology images at various magnifications. This manual will walk you through each lab step-by-step. You will also find Review Sheets in the back of your manual to test your understanding of the key concepts in each lab.

New to Version 4.0

Note to instructors: If you have used previous versions of PhysioEx, here is a summary of what you will find new to version 4.0:

- A new lab on acid-base balance has been added.
- A new Histology Review Supplement has been added, consisting of forty histology slides that relate to topics covered in the PhysioEx lab simulations. In addition, written review worksheets are provided to accompany the slides.
- This lab manual is available in two versions: a CD version containing a CD-ROM of the PhysioEx software, and a web version containing passcodes for accessing PhysioEx online at www.physioex.com. Check the front cover of this manual to see which version you have.

An Instructor's Guide is also available and is free with adoption of this lab manual. Contact your local Benjamin Cummings sales representative for more information. To locate the sales representative for your school, go to **http://www.awl.com/ replocator**

Topics in This Edition

Exercise 1 The Cell—Transport Mechanisms and Cell Permeability: Computer Simulation. Explores how substances cross the cell's membrane. Simple and facilitated diffusion, osmosis, filtration, and active transport are covered.

Exercise 2 Skeletal Muscle Physiology: Computer Simulation. Provides insights into the complex physiology of skeletal muscle. Electrical stimulation, isometric contractions, and isotonic contractions are investigated.

Exercise 3 Neurophysiology of Nerve Impulses: Computer Simulation. Investigates stimuli that elicit action potentials, stimuli that inhibit action potentials, and factors affecting nerve conduction velocity.

Exercise 4 Endocrine System Physiology: Computer Simulation. Investigates the relationship between hormones and metabolism; the effect of estrogen replacement therapy; and the effects of insulin on diabetes.

Exercise 5 Cardiovascular Dynamics: Computer Simulation. Topics of inquiry include vessel resistance and pump (heart) mechanics.

Exercise 6 Frog Cardiovascular Physiology: Computer Simulation. Variables influencing heart activity are examined. Topics include setting up and recording baseline heart activity, the refractory period cardiac muscle, and an investigation of physical and chemical factors that affect enzyme activity.

Exercise 7 Respiratory System Mechanics: Computer Simulation. Investigates physical and chemical aspects of pulmonary function. Students collect data simulating normal lung volumes. Other activities examine factors such as airway resistance and the effect of surfactant on lung function.

Exercise 8 Chemical and Physical Processes of Digestion: Computer Simulation. Turns the student's computer into a virtual chemistry lab where enzymes, reagents, and incubation conditions can be manipulated (in compressed time) to examine factors that affect enzyme activity.

Exercise 9 Renal Physiology: Computer Simulation. Simulates the function of a single nephron. Topics include factors influencing glomerular filtration, the effect of hormones on urine function, and glucose transport maximum.

Exercise 10 Acid-Base Balance: Computer Simulation. Topics include respiratory and metabolic acidosis/alkalosis, as well as renal and respiratory compensation.

Using the Histology Module. Includes over 200 histology images, viewable at various magnifications, with accompanying descriptions and labels.

Getting Started

To use PhysioEx version 4.0, your computer should meet the following minimum requirements.

- **IBM/PC:** Windows 95, 98, NT, 2000, Millennium Edition or higher; Pentium I/266 mHz or faster.
- **Macintosh:** Macintosh 8.6 and above; 604/300 mHz or G3/233 mHz.
- 64 MB RAM (128 MB recommended)
- 800 × 600 screen resolution, millions of colors
- Internet Explorer 5.0 (or higher) *or* Netscape 4.6 (or higher)*
- Flash 6† plug-in
- 4× CD-ROM drive (if using CD-ROM version)
- Printer

*CD version users: Although you do not need a live Internet connection to run the CD, you do need to have a browser installed on your computer. If you do not have a browser, the CD includes a free copy of Netscape which you may install. See instructions on the CD-liner notes.

†CD version users: If you do not have Flash 6 installed on your computer, the CD includes a free Flash installer. See instructions on the CD-liner notes.

Instructions for Getting Started—IBM/PC Users (CD version)

1.Put the PhysioEx CD in the CD-ROM drive. The program should launch automatically. If autorun is disabled on your machine, double click the My Computer icon on your Windows desktop, and then double click the PhysioEx icon.

2.Although you do not need a live Internet connection to run PhysioEx, you do need to have a browser (such as Netscape or Internet Explorer) installed on your computer. If you already have a browser installed, proceed to step 3. If you do not have a browser installed, follow the instructions for installing Netscape found on the liner notes that are packaged with your CD.

3.If you can see a clock onscreen with the clock hands moving, click Proceed. If you cannot see the clock, or if you can see the clock but the hands are not moving, follow the instructions for installing Flash 6 found on the liner notes that are packaged with your CD.

4.On the License Agreement screen, click "Agree" to proceed.

5.On the screen with the PhysioEx icon at the top, click the License Agreement link to read the full agreement. Then close the License Agreement window and click the Main Menu link.

6.From the Main Menu, click on the lab you wish to enter.

Instructions for Getting Started—Mac Users (CD version)

1.Put the PhysioEx CD in the CD-ROM drive. The program should launch automatically. If autorun is disabled on your computer, double click on the PhysioEx icon that appears on your desktop.

2.Although you do not need a live Internet connection to run PhysioEx, you do need to have a browser (such as Netscape or Internet Explorer) installed on your computer. If you already have a browser installed, proceed to step 3. If you do not have a browser installed, follow the instructions for installing Netscape found on the liner notes that are packaged with your CD.

3.If you can see a clock onscreen with the clock hands moving, click Proceed. If you cannot see the clock, or if you can see the clock but the hands are not moving, follow the instructions for installing Flash 6 found on the liner notes that are packaged with your CD.

4.On the License Agreement screen, click "Agree" to proceed.

5.On the screen with the PhysioEx icon at the top, click the License Agreement link to read the full agreement. Then close the License Agreement window and click the Main Menu link.

6.From the Main Menu, click on the lab you wish to enter.

Instructions for Getting Started—Web Users

If you have the web version of this lab manual, follow the instructions for accessing www.physioex.com that appear at the very front of this booklet.

Technical Support

Phone: 800.677.6337
Email: media.support@pearsoned.com
Hours: 8 A.M. to 5 P.M. CST, Monday–Friday

Exercise 1 The Cell—Transport Mechanisms and Permeability: Computer Simulation

Objectives

1. To define *differential permeability; diffusion (simple diffusion, facilitated diffusion,* and *osmosis); isotonic, hypotonic,* and *hypertonic solutions; passive* and *active processes* of transport; *bulk-phase endocytosis; phagocytosis;* and *solute pump.*
2. To describe the processes that account for the movement of substances across the plasma membrane and to indicate the driving force for each.
3. To determine which way substances will move passively through a differentially permeable membrane (given the appropriate information on concentration differences).

The molecular composition of the plasma membrane allows it to be selective about what passes through it. It allows nutrients to enter the cell but keeps out undesirable substances. By the same token, valuable cell proteins and other substances are kept within the cell, and ex-creta or wastes pass to the exterior. This property is known as **differential,** or **selective, permeability.** Transport through the plasma membrane occurs in two basic ways. In **active processes,** the cell provides energy (ATP) to power the transport process. In the other, **passive processes,** the transport process is driven by concentration or pressure differences between the interior and exterior of the cell.

Passive Processes

The two key passive processes of membrane transport are diffusion and filtration. Diffusion is an important transport process for every cell in the body. By contrast, filtration usually occurs only across capillary walls. Each of these will be considered in turn.

Diffusion

Recall that all molecules possess *kinetic energy* and are in constant motion. As molecules move about randomly at high speeds, they collide and ricochet off one another, changing direction with each collision. For a given temperature, all matter has about the same average kinetic energy. Because kinetic energy is directly related to both mass and velocity ($KE = \frac{1}{2} mv^2$), smaller molecules tend to move faster.

When a **concentration gradient** (difference in concentration) exists, the net effect of this random molecular movement is that the molecules eventually become evenly distributed throughout the environment, i.e., the process called diffusion occurs. Hence, **diffusion** is the movement of molecules from a region of their higher concentration to a region of their lower concentration. Diffusion's driving force is the kinetic energy of the molecules themselves.

The diffusion of particles into and out of cells is modified by the plasma membrane, which constitutes a physical barrier. In general, molecules diffuse passively through the plasma membrane if they are small enough to pass through its pores (and are aided by an electrical gradient), or if they can dissolve in the lipid portion of the membrane as in the case of CO_2 and O_2. The diffusion of solute particles dissolved in water through a differentially permeable membrane is called **simple diffusion.** The diffusion of water through a differentially permeable membrane is called *osmosis.* Both simple diffusion and osmosis involve movement of a substance from an area of its higher concentration to one of its lower concentration, i.e., down its concentration gradient.

Solute Transport Through Nonliving Membranes

This computerized simulation provides information on the passage of water and solutes through semipermeable membranes, which may be applied to the study of transport mechanisms in living membrane-bound cells.

Activity 1: Simulating Dialysis (Simple Diffusion)

Choose **Cell Transport Mechanisms and Permeability** from the main menu. The opening screen will appear in a few seconds (Figure 1.1). The primary features on the screen when the program starts are a pair of glass beakers perched atop a solutions dispenser, a dialysis membranes cabinet at the right side of the screen, and a data collection unit at the bottom of the display.

The beakers are joined by a membrane holder, which can be equipped with any of the dialysis membranes from the cabinet. Each membrane is represented by a thin colored line suspended in a gray supporting frame. The solute concentration of dispensed solutions is displayed at the side of each beaker. As you work through the experiments, keep in mind that membranes are three-dimensional; thus what appears as a slender line is actually the edge of a membrane sheet.

The solutions you can dispense are listed beneath each beaker. You can choose more than one solution, and the amount to be dispensed is controlled by clicking (+) to increase concentration, or (−) to decrease concentration. The chosen solutions are then delivered to their beaker by clicking

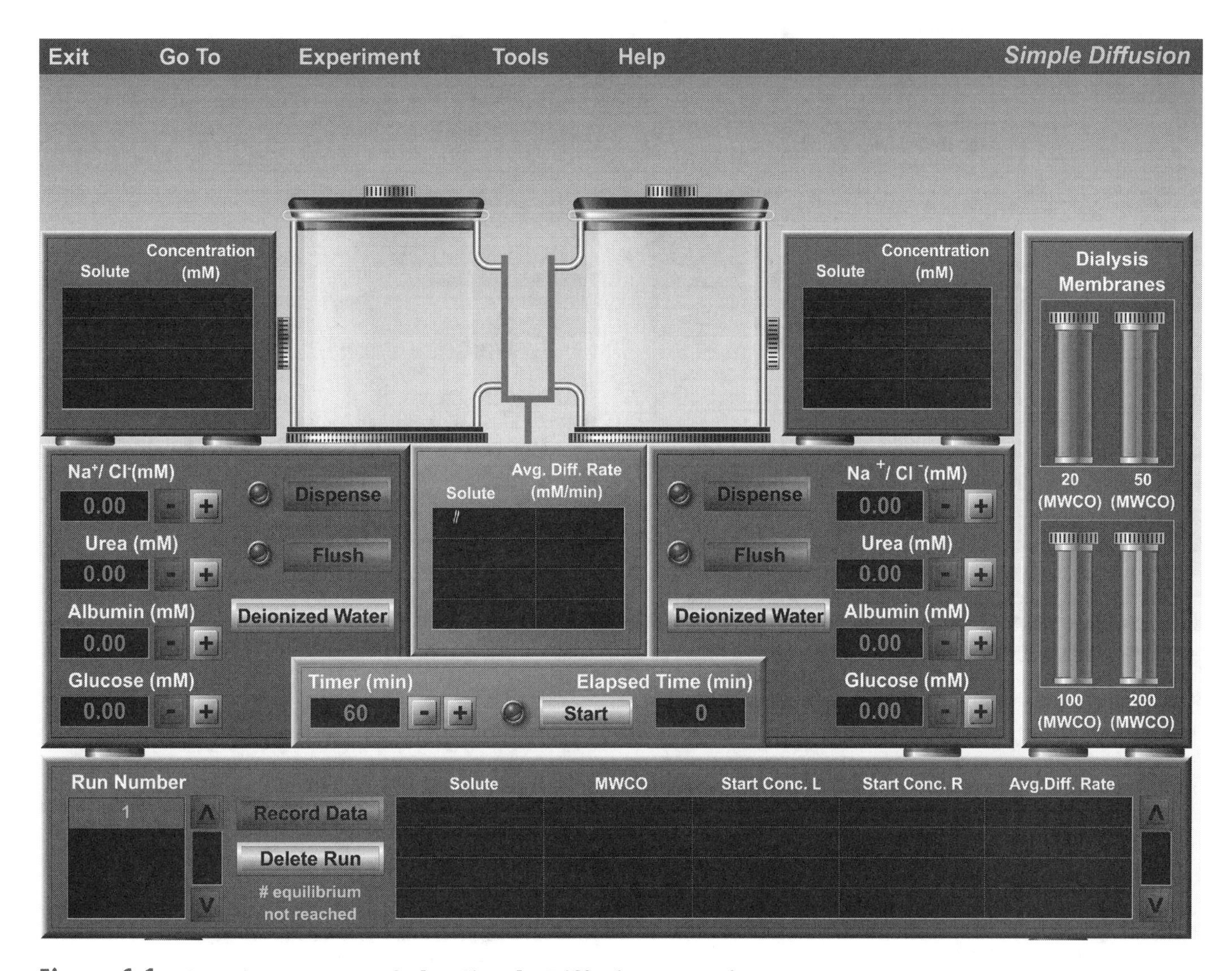

Figure 1.1 Opening screen of the Simple Diffusion experiment.

the **Dispense** button on the same side. Clicking the **Start** button opens the membrane holder and begins the experiment. The **Start** button will become a **Pause** button after it is clicked. To clean the beakers and prepare them for the next run, click **Flush.** Clicking **Pause** and then **Flush** during a run stops the experiment and prepares the beakers for another run. You can adjust the timer for any interval between 5 and 300; the elapsed time is shown in the small window to the right of the timer.

To move dialysis membranes from the cabinet to the membrane holder, click and hold the mouse on the selected membrane, drag it into position between the beakers, and then release the mouse button to drop it into place. Each membrane possesses a different molecular weight cutoff (MWCO), indicated by the number below it. You can think of MWCO in terms of pore size; the larger the MWCO number, the larger the pores in the membrane.

The Run Number window in the data collection unit at the bottom of the screen displays each experimental trial (run). When you click the **Record Data** button, your data is recorded in the computer's memory and is displayed in the data grid at the bottom of the screen. Data displayed in the data grid include the solute (Solute) and membrane (MWCO) used in a run, the starting concentrations in the left and right beakers (Start Conc. L. and Start Conc. R.), and the average diffusion rate (Avg. Dif. Rate). If you are not satisfied with a run, you can click **Delete Run.** Note: Remember NaCl does not move as a molecule. It dissociates to Na^+ and Cl^- ions in water.

1. Click and hold the mouse on the 20 MWCO membrane and drag it to the membrane holder between the beakers. Release the mouse button to lock the membrane into place.

2. Now increase the NaCl concentration to be dispensed by clicking the (+) button under the left beaker until the display window reads 9.00 m*M*. Click **Dispense** to fill the left beaker with 9.00 m*M* NaCl solution.

3. Click the **Deionized Water** button under the right beaker and then click **Dispense** to fill the right beaker with deionized water.

4. Adjust the timer to 60 min (compressed time), then click the **Start** button. When Start is clicked, the barrier between the beakers descends, allowing the solutions in each beaker to have access to the dialysis membrane separating them. Notice that the Start button becomes a Pause button that allows

Chart 1 Dialysis Results

	Membrane (MWCO)			
Solute	20	50	100	200
NaCl				
Urea				
Albumin				
Glucose				

you to momentarily halt the progress of the experiment so you can see instantaneous diffusion or transport rates.

5. Watch the concentration windows at the side of each beaker for any activity. A level above zero in NaCl concentration in the right beaker indicates that Na^+ and Cl^- ions are diffusing from the left into the right beaker through the semipermeable dialysis membrane. Record your results (+ for diffusion, − for no diffusion) in Chart 1. Click the **Record Data** button to keep your data in the computer's memory.

6. Click the 20 MWCO membrane (in the membrane holder) again to automatically return it to the membranes cabinet and then click **Flush** beneath each beaker to prepare for the next run.

7. Drag the next membrane (50 MWCO) to the holder and repeat steps 2 through 6. Continue the runs until you have tested all four membranes. (Remember: click **Flush** beneath each beaker between runs.)

8. Now perform the same experiment for urea, albumin, and glucose by repeating steps 1 through 7 three times. In step 2 you will be dispensing first urea, then albumin, and finally glucose, instead of NaCl.

9. Click **Tools** → **Print Data** to print your data.

Which solute(s) were able to diffuse into the right beaker from the left?

____

Which solute(s) did not diffuse?

____

If the solution in the left beaker contained both urea and albumin, which membrane(s) could you choose to selectively remove the urea from the solution in the left beaker? How would you carry out this experiment?

____

____

____

Assume that the solution in the left beaker contained NaCl in addition to the urea and albumin. How could you set up an experiment so that you removed the urea, but left the NaCl concentration unchanged?

____

____

______________________________ ■

Facilitated Diffusion

Some molecules are lipid insoluble or too large to pass through plasma membrane pores; instead, they pass through the membrane by a passive transport process called **facilitated diffusion.** In this form of transport, solutes combine with carrier protein molecules in the membrane and are then transported *along* or *down* their concentration gradient. Because facilitated diffusion relies on carrier proteins, solute transport varies with the number of available membrane transport proteins.

Activity 2: Simulating Facilitated Diffusion

Click the **Experiment** menu and then choose **Facilitated Diffusion.** The opening screen will appear in a few seconds (Figure 1.2). The basic screen layout is similar to that of the previous experiment with only a few modifications to the equipment. You will notice that only NaCl and glucose solutes are available in this experiment, and you will see a Membrane Builder on the right side of the screen.

The (+) and (−) buttons underneath each beaker adjust solute concentration in the solutions to be delivered into each beaker. Similarly, the buttons in the Membrane Builder allow you to control the number of carrier proteins implanted in the membrane when you click the **Build Membrane** button.

In this experiment, you will investigate how glucose transport is affected by the number of available carrier molecules.

1. The Glucose Carriers window in the Membrane Builder should read 500. If not, adjust to 500 by using the (+) or (−) button.

2. Now click **Build Membrane** to insert 500 glucose carrier proteins into the membrane. You should see the membrane appear as a slender line encased in a support structure within the Membrane Builder. Remember that we are looking at the edge of a three-dimensional membrane.

3. Click on the membrane and hold the mouse button down as you drag the membrane to the membrane holder between the beakers. Release the mouse to lock the membrane into place.

4. Adjust the glucose concentration to be delivered to the left beaker to 2.00 m*M* by clicking the (+) button next to the glucose window until it reads 2.00.

5. To fill the left beaker with the glucose solution, click the **Dispense** button just below the left beaker.

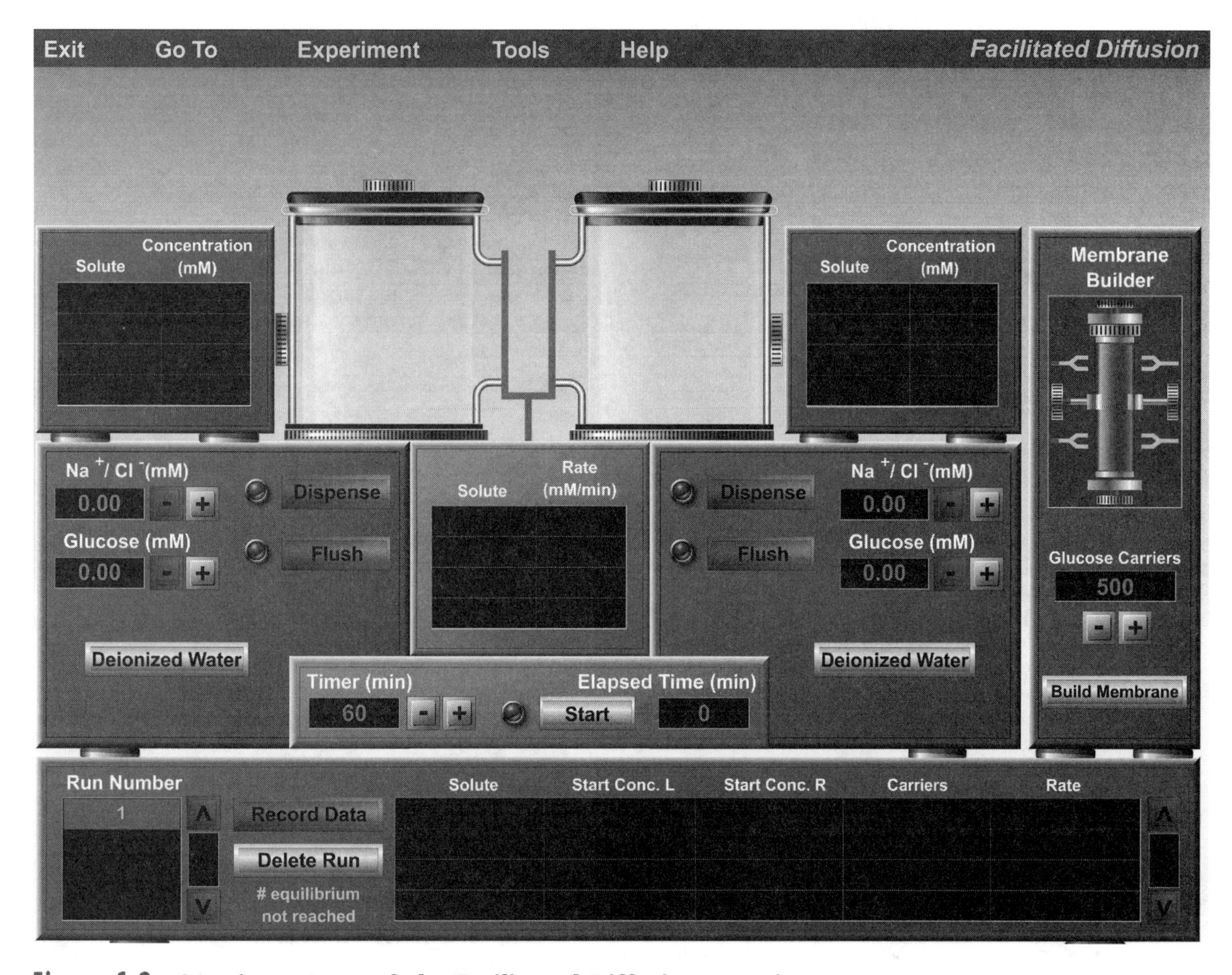

Figure 1.2 Opening screen of the Facilitated Diffusion experiment.

6. Click the **Deionized Water** button below the right beaker, and then click the **Dispense** button. The right beaker will fill with deionized water.

7. Set the timer to 60 min and click **Start.** Watch the concentration windows next to the beakers. When the 60 minutes have elapsed, click the **Record Data** button to display glucose transport rate information in the grid at the lower edge of the screen. Record the glucose transport rate in Chart 2.

Chart 2 Facilitated Diffusion Results

Glucose concentration (mM)	No. of glucose carrier proteins		
	500	700	900
2.00			
8.00			

8. Click the **Flush** button beneath each beaker to remove any residual solution.

9. Click the membrane support to return it to the Membrane Builder. Increase the glucose carriers and repeat steps 2 through 8 using membranes with 700 and then 900 glucose carrier proteins. Record your results in Chart 2 each time.

10. Repeat steps 1 through 9 at 8.00 m*M* glucose concentration. Record your results in Chart 2.

11. Click **Tools** → **Print Data** to print your data.

What happened to the rate of facilitated diffusion as the number of protein carriers increased? Explain your answer.

____

____

____

What do you think would happen to the transport rate if you put the same concentration of glucose into both beakers instead of deionized water in the right beaker?

____

____

Should NaCl have an effect on glucose diffusion? Explain your answer. Use the simulation to see if it does.

____

____

______________________________ ■

Osmosis

A special form of diffusion, the diffusion of water through a semipermeable membrane, is called **osmosis.** Because water can pass through the pores of most membranes, it can move from one side of a membrane to another relatively unimpeded. Osmosis occurs whenever there is a difference in water concentration on the two sides of a membrane.

If we place distilled water on both sides of a membrane, *net* movement of water will not occur; however, water molecules would still move between the two sides of the membrane. In such a situation, we would say that there is no *net* osmosis. The concentration of water in a solution depends on the number of solutes present. Therefore, increasing the solute concentration coincides with a decrease in water concentration. Because water moves down its concentration gradient, it will always move toward the solution with the highest concentration of solutes. Similarly, solutes also move down their concentration gradient. If we position a *fully* permeable membrane (permeable to solutes and water) between two solutions of differing concentrations, then all substances—solutes and water—will diffuse freely, and an equilibrium will be reached between the two sides of the membrane. However, if we use a semipermeable membrane that is impermeable to the solutes, then we have established a condition where water will move but solutes will not. Consequently, water will move toward the more concentrated solution, resulting in a volume increase. By applying this concept to a closed system where volumes cannot change, we can predict that the pressure in the more concentrated solution would rise.

Activity 3:
Simulating Osmotic Pressure

Click the **Experiment** menu and then select **Osmosis.** The opening screen will appear in a few seconds (Figure 1.3). The most notable difference in this experiment screen concerns meters atop the beakers that measure pressure changes in the beaker they serve. As before, (+) and (−) buttons control solute concentrations in the dispensed solutions.

1. Drag the 20 MWCO membrane to the holder between the two beakers.

2. Adjust the NaCl concentration to 8.00 m*M* in the left beaker, and then click the **Dispense** button.

3. Click **Deionized Water** under the right beaker and then click **Dispense.**

4. Set the timer to 60 min and then click **Start** to run the experiment. Pay attention to the pressure displays. Now click the **Record Data** button to retain your data in the computer's memory and also record the osmotic pressure in Chart 3 below.

5. Click the membrane to return it to the membrane cabinet.

6. Repeat steps 1 through 5 with the 50, 100, and 200 MWCO membranes.

Do you see any evidence of pressure changes in either beaker, using any of the four membranes? If so, which ones?

____

Does NaCl appear in the right beaker? If so, which membrane(s) allowed it to pass?

____

____

7. Now perform the same experiment for albumin and glucose by repeating steps 1 through 6. In step 2 you will be dispensing 9.00 m*M* albumin first, and then 10.00 m*M* glucose, instead of NaCl.

8. Click **Tools** → **Print Data** to print your data

Answer the following questions using the results you recorded in Chart 3. Use the simulation if you need help formulating a response.

Chart 3 Osmosis Results (pressure in mm Hg)

	Membrane (MWCO)			
Solute	20	50	100	200
Na^+Cl^-				
Albumin				
Glucose				

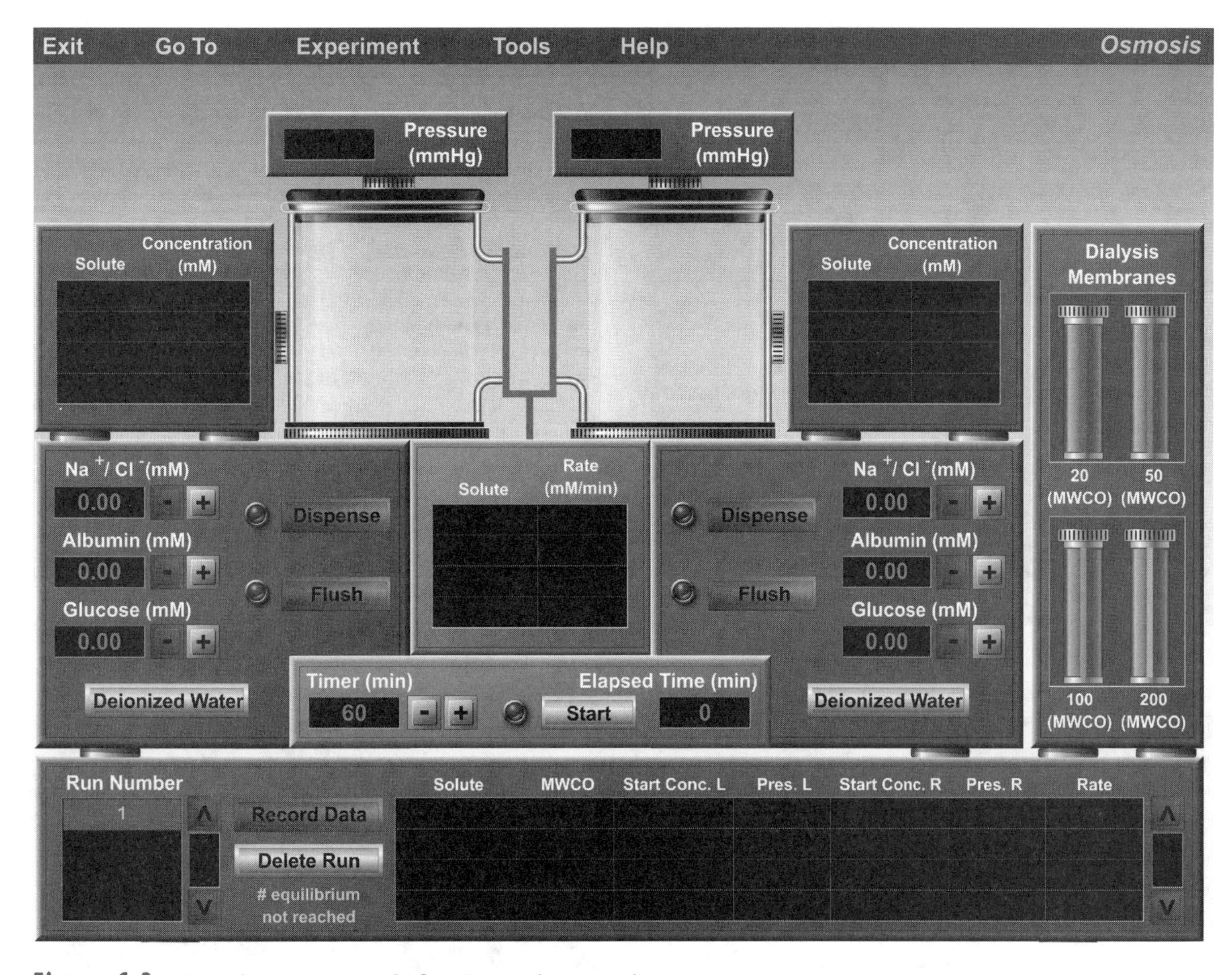

Figure 1.3 Opening screen of the Osmosis experiment.

Explain the relationship between solute concentration and osmotic pressure.

____

____

Will osmotic pressure be generated if solutes are able to diffuse? Explain your answer.

____

____

Because the albumin molecule is much too large to pass through a 100 MWCO membrane, you should have noticed the development of osmotic pressure in the left beaker in the albumin run using the 100 MWCO membrane. What do you think would happen to the osmotic pressure if you replaced the deionized water in the right beaker with 9.00 m*M* albumin in that run? (Both beakers would contain 9.00 m*M* albumin.)

____

What would happen if you doubled the albumin concentration in the left beaker using any membrane?

____

In the albumin run using the 200 MWCO membrane, what would happen to the osmotic pressure if you put 10 m*M* glucose in the right beaker instead of deionized water? Explain your answer.

____

____

____

What if you used the 100 MWCO membrane in the albumin/glucose run described in the previous question?

____

____ ■

Activity 4:
Simulating Filtration

Filtration is the process by which water and solutes pass through a membrane from an area of higher hydrostatic (fluid) pressure into an area of lower hydrostatic pressure. Like diffusion, it is a passive process. For example, fluids and solutes filter out of the capillaries in the kidneys into the kidney tubules because blood pressure in the capillaries is greater than the fluid pressure in the tubules. Filtration is not a selective process. The amount of filtrate—fluids and solutes—formed depends almost entirely on the pressure gradient (the difference in pressure on the two sides of the membrane) and on the size of the membrane pores.

Click the **Experiment** menu and then choose **Filtration.** The opening screen will appear in a few seconds (Figure 1.4). The basic screen elements resemble the other simulations. The top beaker can be pressurized to force fluid through the filtration membrane into the bottom beaker. Any of the filtration membranes can be positioned in the holder between the beakers by drag-and-drop as in the previous experiments. The solutions you can dispense are listed to the right of the top beaker, and are adjusted by clicking the (+) and (−) buttons. The selected solutions are then delivered to the top beaker by clicking **Dispense.** The top beaker is cleaned and prepared for the next run by clicking **Flush.** You can adjust the timer for any interval between 5 and 300; the elapsed time is shown in the window to the right of the timer. When you click the **Record Data** button, your data is recorded in the computer's memory and is displayed in the data grid at the bottom of the screen.

Solute concentrations in the filtrate are automatically monitored by the *Filtrate Analysis Unit* to the right of the bottom beaker. After a run you can detect the presence of any solute remaining on a membrane by using the *Membrane Residue Analysis* unit located above the membrane cabinet.

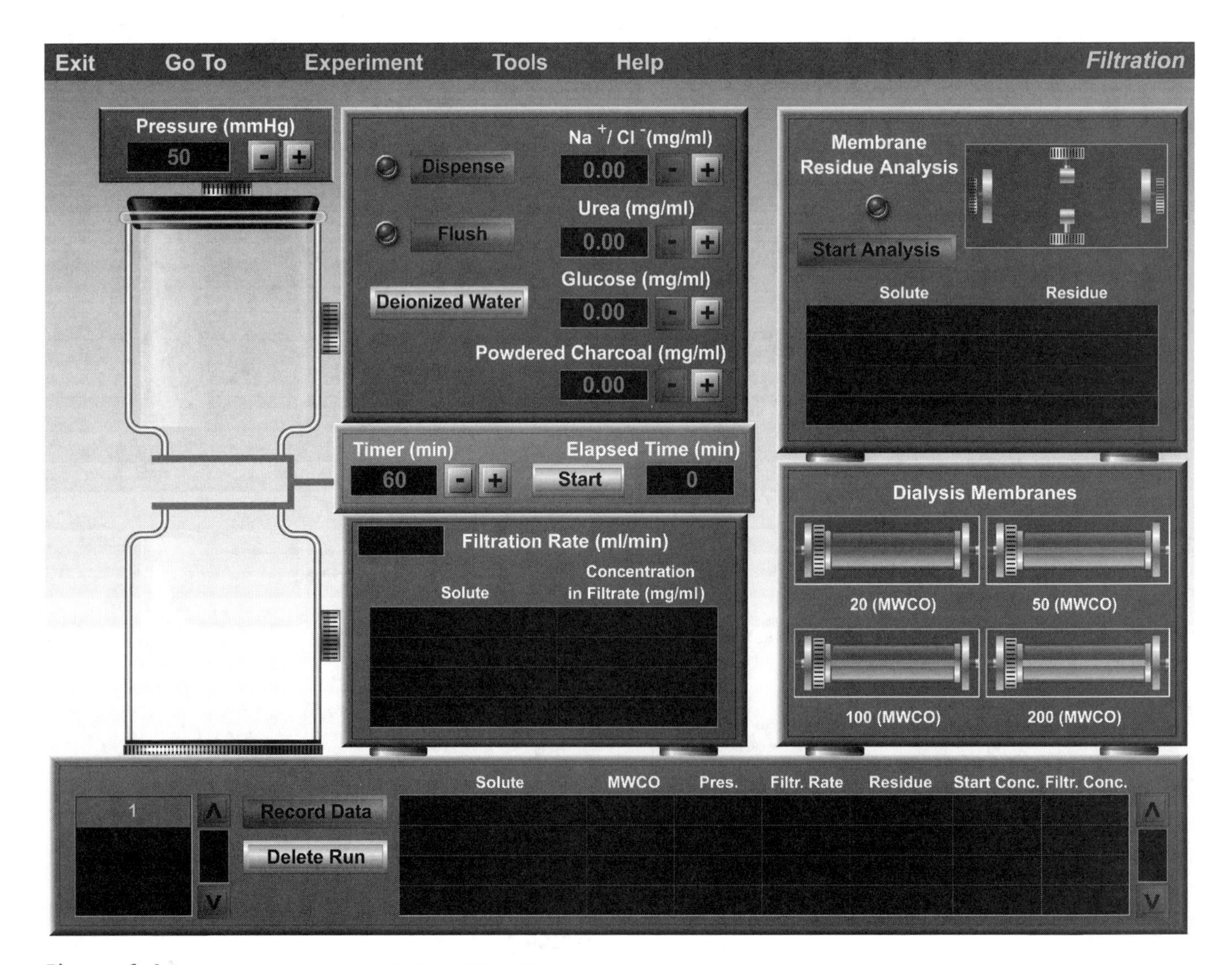

Figure 1.4 Opening screen of the Filtration experiment.

Chart 4 Filtration Results (Filtration Rate, Solute Presence or Absence)

		Membrane (MWCO)			
Solute		20	50	100	200
	Rate				
NaCl	Filtrate				
	Residue				
Urea	Filtrate				
	Residue				
Glucose	Filtrate				
	Residue				
Powdered charcoal	Filtrate				
	Residue				

1. Click and hold the mouse on the 20 MWCO membrane and drag it to the holder below the top beaker. Release the mouse button to lock the membrane into place.

2. Now adjust the NaCl, urea, glucose, and powdered charcoal windows to 5.00 mg/ml each, and then click **Dispense.**

3. If necessary, adjust the pressure unit atop the beaker until its window reads 50 mm Hg.

4. Set the timer to 60 min and then click **Start.** When the Start button is clicked, the membrane holder below the top beaker retracts, and the solution will flow through the membrane into the beaker below.

5. Watch the Filtrate Analysis Unit for any activity. A rise in detected solute concentration indicates that the solute particles are moving through the filtration membrane. At the end of the run, record the amount of solute present in the *filtrate* (mg/ml) and the filtration *rate* in Chart 4.

6. Now drag the 20 MWCO membrane to the holder in the Membrane Residue Analysis unit. Click **Start Analysis** to begin analysis (and cleaning) of the membrane. Record your results for solute *residue* presence on the membrane (+ for present, − for not present) in Chart 4 and click the **Record Data** button to keep your data in the computer's memory.

7. Click the 20 MWCO membrane again to automatically return it to the membranes cabinet and then click **Flush** to prepare for the next run.

8. Repeat steps 1 through 7 using 50, 100, and 200 MWCO membranes.

9. Click **Tools** → **Print Data** to print your data.

Did the membrane's MWCO affect the filtration rate?

____

Which solute did not appear in the filtrate using any of the membranes?

____

What would happen if you increased the driving pressure? Use the simulation to arrive at an answer.

____

Explain how you can increase the filtration rate through living membranes.

____

____

By examining the filtration results, we can predict that the molecular weight of glucose must be:

greater than ________, but less than ________. ■

Active Transport

Whenever a cell expends cellular energy (ATP) to move substances across its membrane, the process is referred to as an *active transport process.* Substances moved across cell membranes by active means are generally unable to pass by diffusion. There are several possible reasons why substances may not be able to pass through a membrane by diffusion: they may be too large to pass through the membrane channels, they may not be lipid soluble, or they may have to move against rather than with a concentration gradient.

In one type of active transport, substances move across the membrane by combining with a protein carrier molecule; the process resembles an enzyme-substrate interaction. ATP provides the driving force, and in many cases the substances move against concentration or electrochemical gradients or both. Some of the substances that are moved into the cells by such carriers, commonly called **solute pumps,** are amino acids and some sugars. Both solutes are lipid-insoluble and too large to pass through the membrane channels, but are necessary for cell life. On the other hand, sodium ions (Na^+) are ejected from the cells by active transport. There is more Na^+ outside the cell than inside, so the Na^+ tends to remain in the cell unless actively transported out. In the body, the most common type of solute pump is the coupled Na^+-K^+ pump that moves Na^+ and K^+ in opposite directions across cellular membranes. 3 Na^+ are ejected for every 2 K^+ entering the cell.

Engulfment processes such as bulk-phase endocytosis and phagocytosis also require ATP. In **bulk-phase endocytosis,** the cell membrane sinks beneath the material to form a small vesicle, which then pinches off into the cell interior. Bulk-phase endocytosis is most common for taking in liquids containing protein or fat.

In **phagocytosis** (cell eating), parts of the plasma membrane and cytoplasm expand and flow around a relatively large or solid material such as bacteria or cell debris and engulf it, forming a membranous sac called a phagosome. The phagosome is then fused with a lysosome and its contents are digested. In the human body, phagocytic cells are mainly found among the white blood cells and macrophages that act as scavengers and help protect the body from disease-causing microorganisms and cancer cells.

You will examine various factors influencing the function of solute pumps in the following experiment.

Activity 5: Simulating Active Transport

Click the **Experiment** menu and then choose **Active Transport.** The opening screen will appear in a few seconds (Figure 1.5). This experiment screen resembles the osmosis experiment screen, except that an ATP dispenser is substituted for the pressure meters atop the beakers. The (+) and (−) buttons control NaCl, KCl, and glucose concentrations in the dispensed solutions. You will use the Membrane Builder to build membranes containing glucose (facilitated diffusion) carrier proteins and active transport Na^+-K^+ pumps.

In this experiment, we will assume that the left beaker represents the cell's interior and the right beaker represents the extracellular space. The Membrane Builder will insert the Na^+-K^+ pumps into the membrane so Na^+ will be pumped toward the right (out of the cell) while K^+ is simultaneously moved to the left (into the cell).

1. In the Membrane Builder, adjust the number of glucose carriers and the number of Na^+-K^+ pumps to 500.
2. Click **Build Membrane,** and then drag the membrane to its position in the membrane holder between the beakers.
3. Adjust the NaCl concentration to be delivered to the left beaker to 9.00 m*M,* then click the **Dispense** button.
4. Adjust the KCl concentration to be delivered to the right beaker to 6.00 m*M,* then click **Dispense.**
5. Adjust the ATP dispenser to 1.00 m*M*, then click **Dispense ATP.** This action delivers the chosen ATP concentration to both sides of the membrane.
6. Adjust the timer to 60 min, and then click **Start.** Click **Record Data** after each run.
7. Click **Tools** → **Print Data** to print your data.

Watch the solute concentration windows at the side of each beaker for any changes in Na^+ and K^+ concentrations. The Na^+ transport rate slows and then stops before transport has completed. Why do you think that this happens?

____

What would happen if you did not dispense any ATP?

____

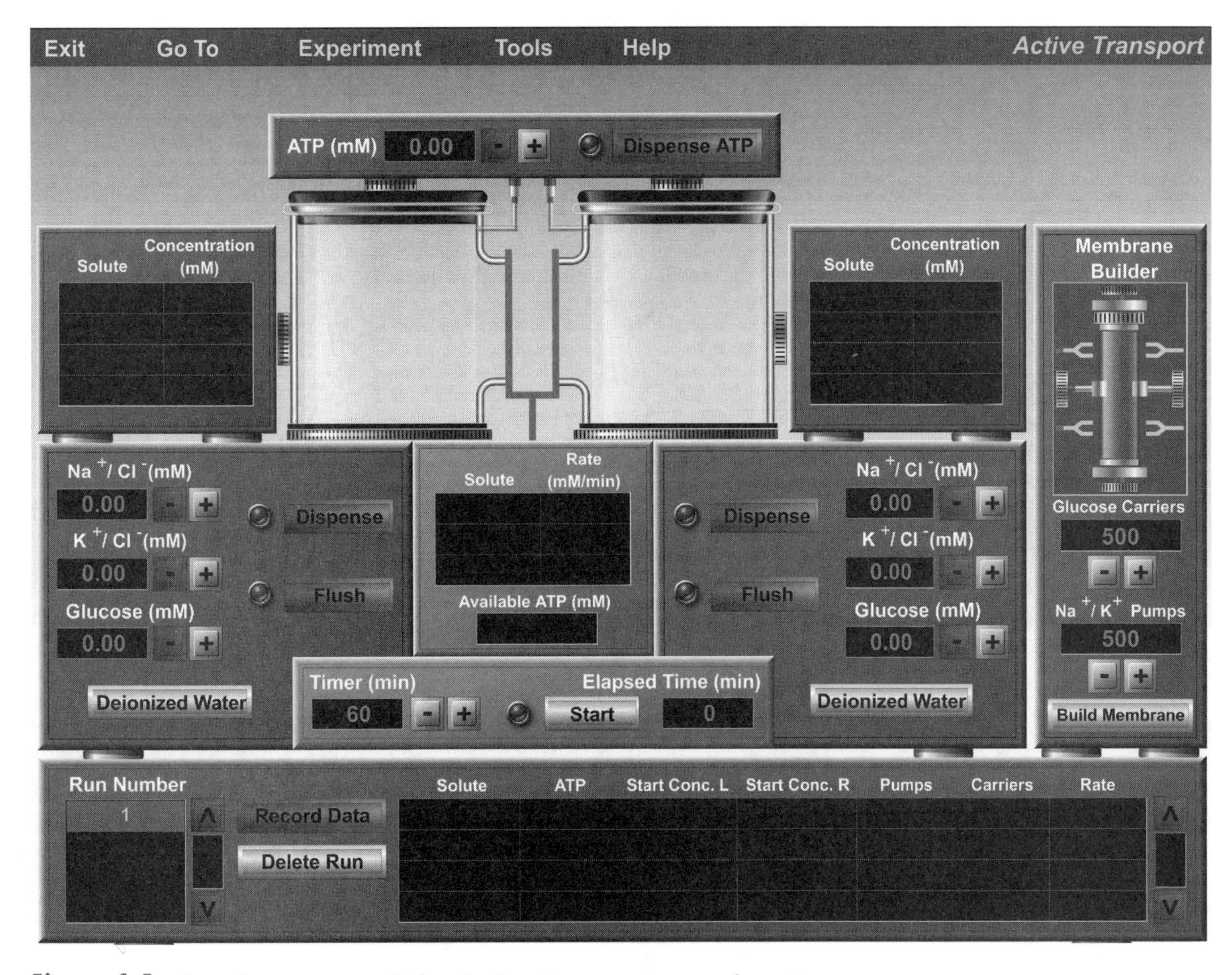

Figure 1.5 Opening screen of the Active Transport experiment.

7. Click either Flush button to clean both beakers. Repeat steps 3 through 6, adjusting the ATP concentration to 3.00 m*M* in step 5.

Has the amount of Na^+ transported changed?

____

Do these results support your ideas in step 6 above? ______

What would happen if you decreased the number of Na^+-K^+ pumps?

____

Explain how you could show that this phenomenon is not just simple diffusion. (Hint: adjust the Na^+ concentration in the right beaker.)

____

____

8. Now repeat steps 1 through 6 dispensing 9.00 m*M* NaCl into the left beaker and 10.00 m*M* NaCl into the right beaker (instead of 6.00 m*M* KCl). Is Na^+ transport affected by this change? Explain your answer.

____

____

What would happen to the rate of ion transport if we increased the number of Na^+-K^+ pump proteins?

___

Would Na^+- and K^+ transport change if we added glucose solution?

___

9. Click **Tools** → **Print Data** to print your data.

Try adjusting various membrane and solute conditions and attempt to predict the outcome of experimental trials. For example, you could dispense 10 m*M* glucose into the right beaker instead of deionized water. ■

Skeletal Muscle Physiology: Computer Simulation

Exercise 2

Objectives

1. To define these terms used in describing muscle physiology: *multiple motor unit summation, maximal stimulus, treppe, wave summation, tetanus.*
2. To identify two ways that the mode of stimulation can affect muscle force production.
3. To plot a graph relating stimulus strength and twitch force to illustrate graded muscle response.
4. To explain how slow, smooth, sustained contraction is possible in a skeletal muscle.
5. To graphically understand the relationships between passive, active, and total forces.
6. To identify the conditions under which muscle contraction is isometric or isotonic.
7. To describe in terms of length and force the transitions between isometric and isotonic conditions during a single muscle twitch.
8. To describe the effects of resistance and starting length on the initial velocity of shortening.
9. To explain why muscle force remains constant during isotonic shortening.
10. To explain experimental results in terms of muscle structure.

Skeletal muscles are composed of hundreds to thousands of individual cells, each doing their share of work in the production of force. As their name suggests, skeletal muscles move the skeleton. Skeletal muscles are remarkable machines; while allowing us the manual dexterity to create magnificent works of art, they are also capable of generating the brute force needed to lift a 100-lb sack of concrete. When a skeletal muscle from an experimental animal is electrically stimulated, it behaves in the same way as a stimulated muscle in the intact body, that is, *in vivo.* Hence, such an experiment gives us valuable insight into muscle behavior.

This set of computer simulations demonstrates many important physiological concepts of skeletal muscle contraction. The program graphically provides all the equipment and materials necessary for you, the investigator, to set up experimental conditions and observe the results. In student-conducted laboratory investigations there are many ways to approach a problem, and the same is true of these simulations. The instructions will guide you in your investigation, but you should also try out alternate approaches to gain insight into the logical methods used in scientific experimentation.

Try this approach: As you work through the simulations for the first time, follow the instructions closely and answer the questions posed as you go. Then try asking "What if . . . ?" questions to test the validity of your theories. The major advantages of these computer simulations are that the muscle cannot be accidentally damaged, lab equipment will not break down at the worst possible time, and you will have ample time to think critically about the processes being investigated.

Because you will be working with a simulated muscle and oscilloscope display, you need to watch both carefully during the experiments. Think about what is happening in each situation. You need to understand how you are experimentally manipulating the muscle in order to understand your results.

Electrical Stimulation

A contracting skeletal muscle will produce force and/or shortening when nervous or electrical stimulation is applied. The graded contractile response of a whole muscle reflects the number of motor units firing at a given time. Strong muscle contraction implies that many motor units are activated and each unit has maximally contracted. Weak contraction means that few motor units are active; however, the activated units are maximally contracted. By increasing the number of motor units firing, we can produce a steady increase in muscle force, a process called recruitment or motor unit summation.

Regardless of the number of motor units activated, a single contraction of skeletal muscle is called a muscle twitch. A tracing of a muscle twitch is divided into three phases: latent, contraction, and relaxation. The latent phase is a short period between the time of stimulation and the beginning of contraction. Although no force is generated during this interval, chemical changes occur intracellularly in preparation for contraction. During contraction, the myofilaments are sliding past each other and the muscle shortens. Relaxation takes place when contraction has ended and the muscle returns to its normal resting state and length.

The first activity you will conduct simulates an isometric, or fixed length, contraction of an isolated skeletal muscle. This activity allows you to investigate how the strength and frequency of an electrical stimulus affect whole muscle function. Note that these simulations involve indirect stimulation by an electrode placed on the surface of the muscle. This differs from the situation *in vivo* where each fiber in the muscle receives direct stimulation via a nerve ending. In other words, increasing the intensity of the electrical stimulation mimics how the nervous system increases the number of motor units activated.

Single Stimulus

Choose **Skeletal Muscle Physiology** from the main menu. The opening screen will appear in a few seconds (Figure 2.1). The oscilloscope display, the grid at the top of the screen, is the most important part of the screen because it graphically displays the contraction data for analysis. Time is displayed on the horizontal axis. A full sweep is initially set at 200

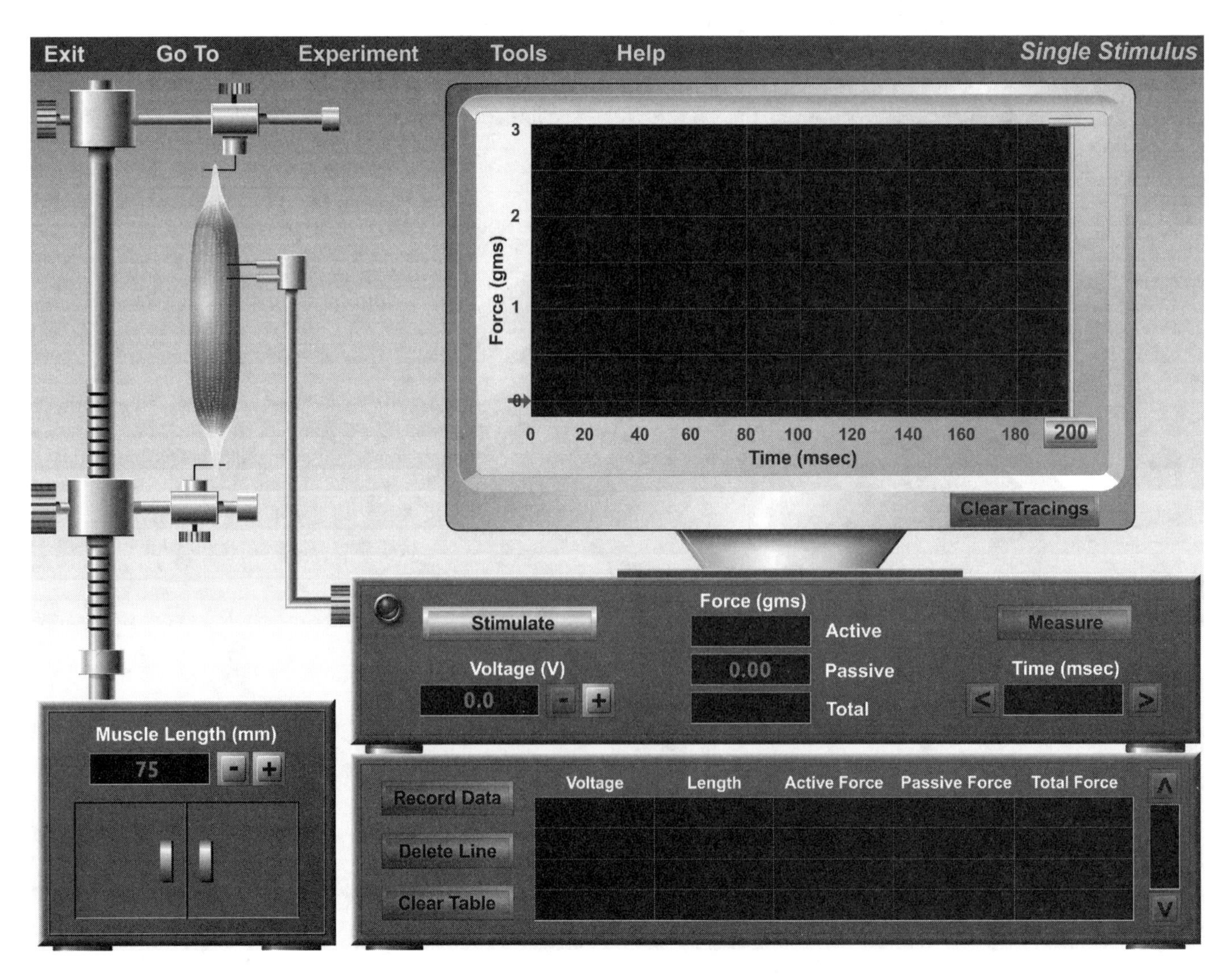

Figure 2.1 Opening screen of the Single Stimulus experiment.

msec. However, you can adjust the sweep time from 200 msec to 1000 msec by clicking and dragging the **200** msec button at the lower right corner of the oscilloscope display to the left to a new position on the time axis. The force (in grams) produced by muscle contraction is displayed on the vertical axis. Clicking the **Clear Tracings** button erases all muscle twitch tracings from the oscilloscope display.

The *electrical stimulator* is the equipment seen just beneath the oscilloscope display. Clicking **Stimulate** delivers the electrical shock to the muscle through the electrodes lying on the surface of the muscle. Stimulus voltage is set by clicking the (+) or (−) buttons next to the voltage window. Three small windows to the right of the Stimulate button display the force measurements. *Active force* is produced during muscle contraction, while *passive force* results from the muscle being stretched (much like a rubber band). The *total force* is the sum of active and passive forces. The red arrow at the left of the oscilloscope display is an indicator of passive force. After the muscle is stimulated, the **Measure** button at the right edge of the electrical stimulator becomes active. When the Measure button is clicked, a vertical orange line will be displayed at the left edge of the oscilloscope display. Clicking the arrow buttons below the Measure button moves the orange line horizontally across the screen. The Time window displays the difference in time between the zero point on the X-axis and the intersection between the orange measure line and the muscle twitch tracing.

The muscle is suspended in the support stand to the left of the oscilloscope display. The hook through the upper tendon of the muscle is part of the force transducer, which measures the force produced by the muscle. The hook through the lower tendon secures the muscle in place. The weight cabinet just below the muscle support stand is not active in this experiment; it contains weights you will use in the isotonic contraction part of the simulation. You can adjust the starting length of the muscle by clicking the (+) or (−) buttons located next to the Muscle Length display window.

When you click the **Record Data** button in the data collection unit below the electrical stimulator, your data is recorded in the computer's memory and is displayed in the data grid at the bottom of the screen. Data displayed in the data grid include the voltage, muscle length, and active, passive, and total force measurements. If you are not satisfied with a single run, you can click **Delete Line** to erase a single line of data. Clicking the **Clear Table** button will remove all accumulated data in the experiment and allow you to start over.

Activity 1:
Practicing Generating a Tracing

1. Click the **Stimulate** button once. Because the beginning voltage is set to zero, no muscle activity should result. You will see a blue line moving across the bottom of the oscilloscope display. This blue line will indicate muscle force in the experiments. If the tracings move too slowly across the screen, click and hold the **200** button at the lower right corner of the oscilloscope and drag it to the left to the 40 msec mark and release it. This action resets the total sweep time to 1000 msec to speed up the display time.

2. Click and hold the (+) button beneath the Stimulate button until the voltage window reads 3.0 volts. Click **Stimulate** once. You will see the muscle react, and a contraction tracing will appear on the screen. Notice that the muscle tracing color alternates between blue and yellow each time the Stimulate button is clicked to enhance the visual difference between twitch tracings. You can click the **Clear Tracings** button as needed to clean up the oscilloscope display. To retain your data, click the **Record Data** button at the end of each stimulus.

3. Change the voltage to 5.0 volts and click **Stimulate** again. Notice how the force of contraction also changes. Identify the latent, contraction, and relaxation phases in the tracings.

4. You may print your data by clicking **Tools** → **Print Data**. You may print out hard copies of the graphs you generate by clicking on **Tools** and then selecting **Print Graph.**

Feel free to experiment with anything that comes to mind to get a sense of how whole muscle responds to an electrical stimulus. ■

Activity 2:
Determining the Latent Period

1. Click **Clear Tracings** to erase the oscilloscope display. The voltage should be set to 5.0 volts.

2. Drag the **200** msec button to the right edge of the oscilloscope.

3. Click the **Stimulate** button once and allow the tracing to complete.

4. When you measure the length of the latent period from a printed graph, you measure the time between the application of the stimulus and the beginning of the first observable response (increase in force). The computer can't "look ahead," anticipating the change in active force. To measure the length of the latent period using the computer, click the **Measure** button. Then click the right arrow button next to the **Time** window repeatedly until you notice the first increase in the Active Force window. This takes you beyond the actual length of the latent period. Now click the left arrow button next to the **Time** window until the Active Force window again reads zero. At this point the computer is measuring the time between the application of the stimulus and the last point where the active force is zero (just prior to contraction).

How long is the latent period? ______________ msec

What occurs in the muscle during this apparent lack of activity?

____

____ ■

The Graded Muscle Response to Increased Stimulus Intensity

As the stimulus to a muscle is increased, the amount of force produced by the muscle also increases. As more voltage is delivered to the whole muscle, more muscle fibers are activated and the total force produced by the muscle is increased. Maximal contraction occurs when all the muscle cells have been activated. Any stimulation beyond this voltage will not increase the force of contraction. This experiment mimics muscle activity *in vivo* where the recruitment of additional motor units increases the total force produced. This phenomenon is called *multiple motor unit summation.*

Activity 3:
Investigating Graded Muscle Response to Increased Stimulus Intensity

1. Click **Clear Tracings** if there are tracings on your screen.

2. Set the voltage to 0.0 and click **Stimulate.**

3. Click **Record Data.** If you decide to redo a single stimulus, choose the data line in the grid and click **Delete Line** to erase that single line of data. If you want to repeat the entire experiment, click the **Clear Table** button to erase all data recorded to that point.

4. Repeat steps 2 and 3, increasing the voltage by 0.5 each time until you reach the maximum voltage of 10.0. Be sure to select **Record Data** each time.

5. Observe the twitch tracings. Click on the **Tools** menu and then choose **Plot Data.**

6. Use the slider bars to display Active Force on the Y-axis and Voltage on the X-axis.

7. Use your graph to answer the following questions:

What is the minimal or threshold stimulus? _______ V

What is the maximal stimulus? _______ V

How can you explain the increase in force that you observe?______________________________

____

What type of summation is this? ______________

____

8. Click **Print Plot** at the top left corner of the Plot Data window to print a hard copy of the graph. When finished, click the "x" box at the top right of the plot window.

9. Click **Tools** → **Print Data** to print your data. ■

Multiple Stimulus

Choose **Multiple Stimulus** from the **Experiment** menu. The opening screen will appear in a few seconds (Figure 2.2).

The only significant change to the on-screen equipment is found in the electrical stimulator. The measuring equipment has been removed and other controls have been added: The **Multiple Stimulus** button is a toggle that allows you to alternately start and stop the electrical stimulator. When Multiple Stimulus is first clicked, its name changes to Stop Stimulus and electrical stimuli are delivered to the muscle at the rate specified in the Stimuli/sec window until the muscle completely fatigues or the stimulator is turned off. The stimulator is turned off by clicking the **Stop Stimulus** button. The stimulus rate is adjusted by clicking the (+) or (−) buttons next to the stimuli/sec window.

Activity 4:
Investigating Treppe

When a muscle first contracts, the force it is able to produce is less than the force it is able to produce in subsequent contractions within a relatively narrow time span. A myogram, a recording of a muscle twitch, reveals this phenomenon as the **treppe,** or staircase, effect. For the first few twitches, each successive stimulation produces slightly more force than the previous contraction as long as the muscle is allowed to fully relax between stimuli, and the stimuli are delivered relatively close together. Treppe is thought to be caused by increased efficiency of the enzyme systems within the cell and increased availability of intracellular calcium.

1. The voltage should be set to 8.2 volts and the muscle length should be 75 mm.

2. Drag the **200** msec button to the center of the X-axis time range.

3. Be sure that you fully understand the following three steps before you proceed.

- Click **Single Stimulus.** Watch the twitch tracing carefully.
- After the tracing shows that the muscle has completely relaxed, immediately click **Single Stimulus** again.
- When the second twitch completes, click **Single Stimulus** once more and allow the tracing to complete.

What happens to force production with each subsequent stimulus?

____

______________________________ ■

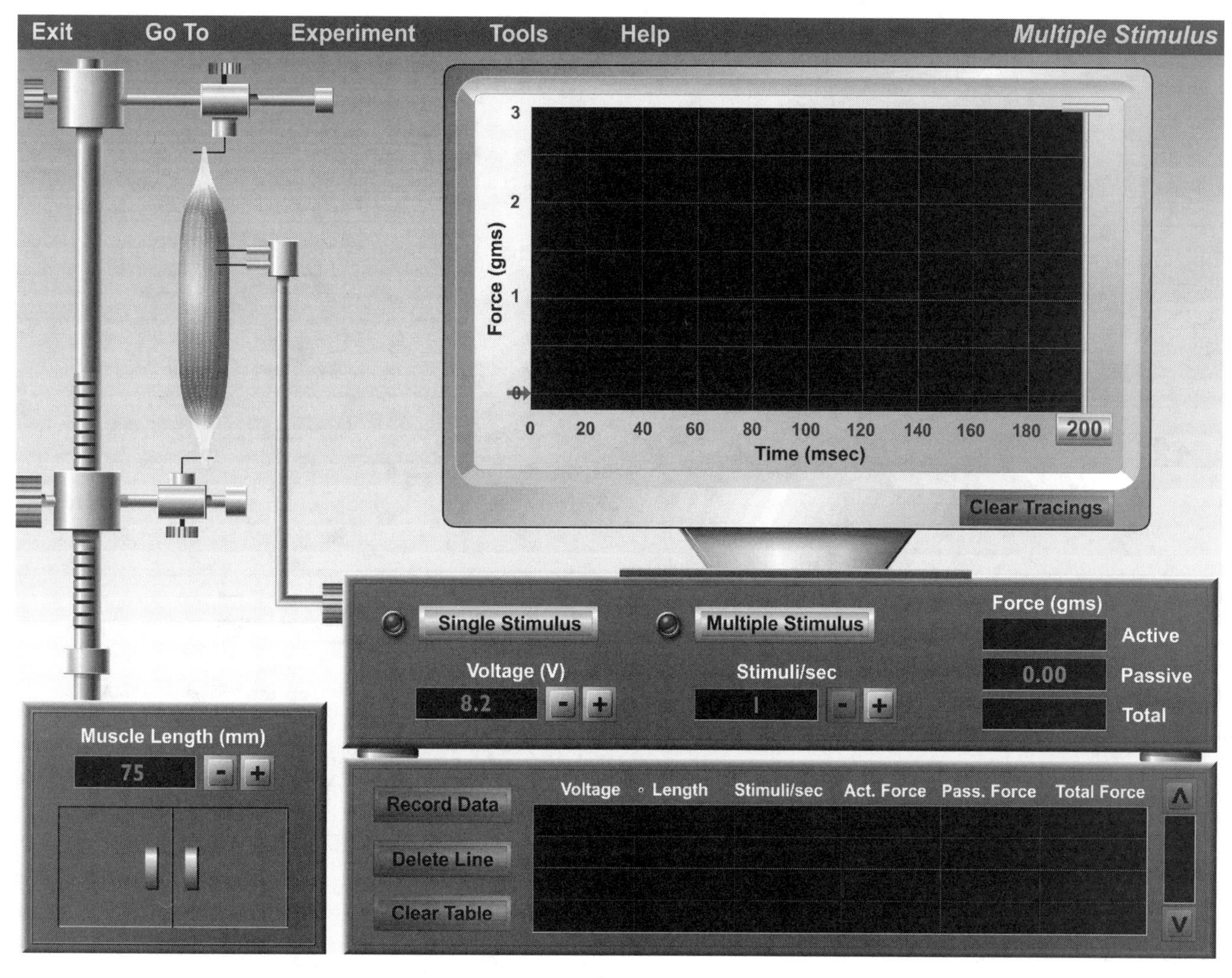

Figure 2.2 Opening screen of the Multiple Stimulus experiment.

Activity 5: Investigating Wave Summation

As demonstrated in Activity 3 with single stimuli, multiple motor unit summation is one way to increase the amount of force produced by muscle. Multiple motor unit summation relied on increased stimulus *intensity* in that simulation. Another way to increase force is by wave, or temporal, summation. **Wave summation** is achieved by increasing the stimulus *frequency,* or rate of stimulus delivery to the muscle. Wave summation occurs because the muscle is already in a partially contracted state when subsequent stimuli are delivered.

Tetanus can be considered an extreme form of wave summation that results in a steady, sustained contraction. In effect, the muscle does not have any chance to relax because it is being stimulated at such a high frequency. This "fuses" the force peaks so that we observe a smooth tracing.

1. Click **Clear Tracings** to erase the oscilloscope display.

2. Set and keep the voltage at the maximal stimulus (8.2 volts) and the muscle length at 75 mm.

3. Drag the **200** msec button to the right edge of the oscilloscope display unless you are using a slower computer.

4. Click **Single Stimulus,** and then click **Single Stimulus** again when the muscle has relaxed about halfway.

You may click **Tools** → **Print Graph** any time you wish to print graphs generated during this activity.

Is the peak force produced in the second contraction greater than that produced by the first stimulus?

____

5. Try stimulating again at greater frequencies by clicking the Single Stimulus button several times in rapid succession.

Is the total force production even greater? ________

6. To see if you can produce smooth, sustained contraction at Active Force = 2 gms, try rapidly clicking **Single Stimulus** several times.

Is it possible to produce a smooth contraction (fused force peaks), or does the force rise and fall periodically?

____

7. Using the same method as in step 6, try to produce a smooth contraction at 3 gms.

Is the trace smoother (less height between peaks and valleys) this time?

____

Because there is a limit to how fast you can manually click the Single Stimulus button, what do you think would happen to the smoothness of the tracing if you could click even faster?

____

____

____

8. So far you have been using the maximal stimulus. What do you think would happen if you used a lower voltage?

____

____

____

____

9. Use the concepts of motor unit summation and stimulus frequency to explain how human skeletal muscles work to achieve smooth, steady contractions at all desired levels of force.

____

____

____

____ ■

Activity 6:
Investigating Fusion Frequency

1. Click **Clear Tracings** to erase the oscilloscope display.

2. The voltage should be set to 8.2 volts and the muscle length should be 75 mm.

3. Adjust the stimulus rate to 30 stimuli/sec.

4. The following steps constitute a single "run." Become familiar with the procedure for completing a run before continuing.

- Click **Multiple Stimulus.**
- When the tracing is close to the right side of the screen, click **Stop Stimulus** to turn off the stimulator.
- Click **Record Data** to retain your data in the grid in the bottom of the screen and the computer's memory. Click **Tools → Print Graph** to print a hard copy of your graph.

If you decide to redo a single run, choose the data line in the grid and click **Delete Line** to erase that single line of data. If you want to repeat the entire experiment, click the **Clear Table** button to erase all data recorded thus far.

Describe the appearance of the tracing.

____

____

5. Repeat step 4, increasing the stimulation rate by 10 stimuli/sec each time up to 150 stimuli/sec.

How do the tracings change as the stimulus rate is increased?

____

6. When you have finished observing the twitch tracings, click the **Tools** menu and then choose **Plot Data.**

7. Set the Y-axis slider to display Active Force and the X-axis slider to display Stimuli/sec.

From your graph, estimate the stimulus rate above which there appears to be no significant increase in force.

_______________ stimuli/sec

This rate is the fusion frequency, also called tetanus.

8. Click **Print Plot** at the top left of the Plot Data window. When finished, click the "x" box at the top right of the plot window.

9. Reset the stimulus rate to the fusion frequency.

10. Try to produce a smooth contraction at Force = 2 gms and Force = 3 gms by adjusting only the stimulus intensity, or voltage, using the following procedure.

- Decrease the voltage to a starting point of 1.0 volt and click **Multiple Stimulus.**
- Click **Stop Stimulus** to turn off the stimulator when the tracing is near the right side of the oscilloscope display.
- If the force produced is not smooth and continuous at the desired level of force, increase the voltage in 0.1-volt increments and stimulate as above until you achieve a smooth force at 2 gms and again at 3 gms.

What stimulus intensity produced smooth force at Force = 2 gms?

_______________ V

Which intensity produced smooth contraction at Force = 3 gms?

_______________ V

Explain what must be happening in the muscle to achieve smooth contraction at different force levels.

■

Activity 7:

Investigating Muscle Fatigue

A prolonged period of sustained contraction will result in muscle fatigue, a condition in which the tissue has lost its ability to contract. Fatigue results when a muscle cell's ATP consumption is faster than its production. Consequently, increasingly fewer ATP molecules are available for the contractile parts within the muscle cell.

1. Click **Clear Tracings** to erase the oscilloscope display.
2. The voltage should be set to 8.2 volts and the muscle length should be 75 mm.
3. Adjust the stimulus rate to 120 stimuli/sec.
4. Click **Multiple Stimulus** and allow the tracing to sweep through three screens and then click **Stop Stimulus** to stop the stimulator.

Click **Tools** → **Print Graph** at any time to print graphs.

Why does the force begin to fall with time? Note that a fall in force indicates muscle fatigue.

5. Click **Clear Tracings** to erase the oscilloscope display. Keep the same settings as before.
6. You will be clicking **Multiple Stimulus** on and off three times to demonstrate fatigue with recovery. Read the steps below before proceeding.

- Click **Multiple Stimulus.**
- When the tracing reaches the middle of the screen, briefly turn off the stimulator by clicking **Stop Stimulus,** then immediately click **Multiple Stimulus** again.
- You will see a dip in the force tracing where you turned the stimulator off and then on again. The force tracing will continue to drop as the muscle fatigues.
- Before the muscle fatigues completely, repeat the on/off cycle twice more without clearing the screen.

Turning the stimulator off allows a small measure of recovery. The muscle will produce force for a longer period if the stimulator is briefly turned off than if the stimulations were allowed to continue without interruption. Explain why.

7. To see the difference between continuous multiple stimulation and multiple stimulation with recovery, click **Multiple Stimulus** and let the tracing fall without interruption to zero force. This tracing will follow the original myogram exactly until the first "dip" is encountered, after which you will notice a difference in the amount of force produced between the two runs.

Describe the difference between the current tracing and the myogram generated in step 6.

■

Isometric Contraction

Isometric contraction is the condition in which muscle length does not change regardless of the amount of force generated by the muscle (*iso* = same, *metric* = length). This is accomplished experimentally by keeping both ends of the muscle in a fixed position while stimulating it electrically. Resting length (length of the muscle before contraction) is an important factor in determining the amount of force that a muscle can develop. Passive force is generated by stretching the muscle, and is due to the elastic properties of the tissue itself. Active force is generated by the physiological contraction of the muscle. Think of the muscle as having two force properties: it exerts passive force when it is stretched (like a rubber band exerts passive force), and active force when it contracts. Total force is the sum of passive and active forces, and it is what we experimentally measure.

This simulation allows you to set the resting length of the experimental muscle and stimulate it with individual maximal stimulus shocks. A graph relating the forces generated to the length of the muscle will be automatically plotted as you stimulate the muscle. The results of this simulation can then be applied to human muscles in order to understand how

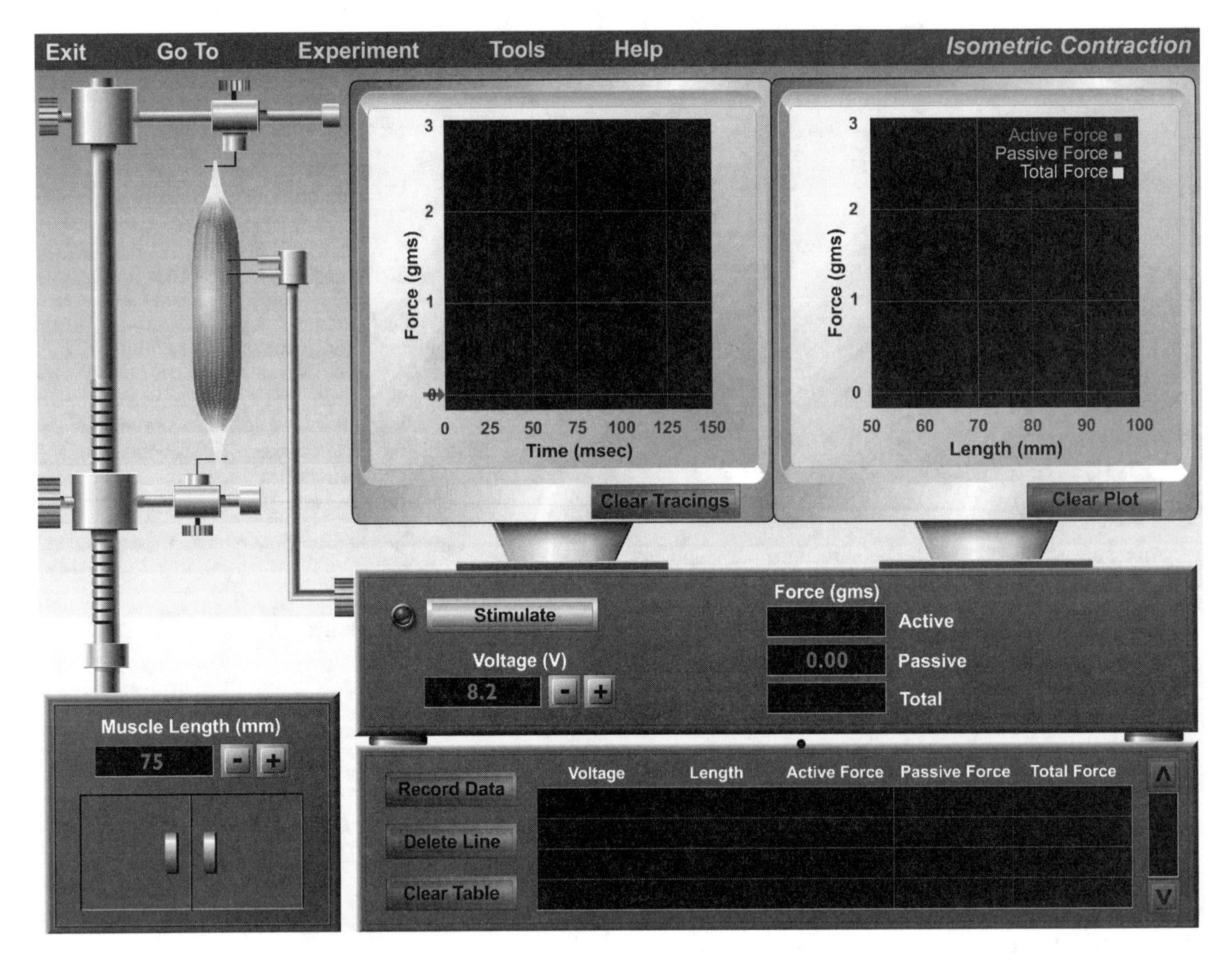

Figure 2.3 Opening screen of the Isometric Contraction experiment.

optimum resting length will result in maximum force production. In order to understand why muscle tissue behaves as it does, it is necessary to comprehend contraction at the cellular level. Hint: If you have difficulty understanding the results of this exercise, review the sliding filament model of muscle contraction. Then think in terms of sarcomeres that are too short, too long, or just the right length.

Choose **Isometric Contraction** from the **Experiment** menu. The opening screen will appear in a few seconds (Figure 2.3). Notice that the oscilloscope is now divided into two parts. The left side of the scope displays the muscle twitch tracing. The active, passive, and total force data points are plotted on the right side of the screen.

Activity 8: Investigating Isometric Contraction

1. The voltage should be set to the maximal stimulus (8.2 volts) and the muscle length should be 75 mm.

2. To see how the equipment works, stimulate once by clicking **Stimulate.** You should see a single muscle twitch tracing on the left oscilloscope display, and three data points representing active, passive, and total force on the right display. The yellow box represents the total force and the red dot it contains symbolizes the superimposed active force. The green square represents the passive force data point.

3. Try adjusting the muscle length by clicking the (+) or (−) buttons located next to the Muscle Length window and watch the effect on the muscle.

4. When you feel comfortable with the equipment, click **Clear Tracings** and **Clear Plot.**

5. Now stimulate at different muscle lengths using the following procedure.

- Shorten the muscle to a length of 50 mm by clicking the (−) button next to the Muscle Length window.
- Click **Stimulate** and when the tracing is complete, click **Record Data.**
- Repeat the **Stimulate** and **Record Data** sequence, increasing the muscle length by 2 mm each time until you reach the maximum muscle length of 100 mm.

6. Carefully examine the active, passive, and total force plots in the right oscilloscope display.
7. Click **Tools** → **Print Data** to print your data.

What happens to the passive and active forces as the muscle length is increased from 50 mm to 100 mm?

Passive force:

____

____

Active force:

____

____

Total force:

____

____

Explain the dip in the total force curve. (Hint: keep in mind you are measuring the sum of active and passive forces.)

____

____

____

____

______________________________ ■

Isotonic Contraction

During isotonic contraction, muscle length changes, but the force produced stays the same (*iso* = same, *tonic* = force). Unlike the isometric exercise in which both ends of the muscle are held in a fixed position, one end of the muscle remains free in the isotonic contraction exercise. Different weights can then be attached to the free end while the other end is fixed in position on the force transducer. If the weight is not too great, the muscle will be able to lift it with a certain velocity. You can think of lifting an object from the floor as an example: if the object is light it can be lifted quickly (high velocity), whereas a heavier weight will be lifted with a slower velocity. Try to transfer the idea of what is happening in the simulation to the muscles of your arm when you lift a weight. The two important variables in this exercise are starting length of the muscle and resistance (weight) applied. You have already examined the effect of starting length on muscle force production in the previous exercise. Now you will change both muscle length and resistance to investigate how such changes affect the speed of skeletal muscle shortening. Both variables can be independently altered and the results are graphically presented on the screen.

Choose **Isotonic Contraction** from the **Experiment** menu. The opening screen will appear in a few seconds (Figure 2.4). The general operation of the equipment is the same as in the previous experiments. In this simulation, the weight cabinet doors are open. You will attach weights to the lower tendon of the muscle by clicking and holding the mouse on any weight in the cabinet and then dragging-and-dropping the weight's hook onto the lower tendon. The Muscle Length window displays the length achieved when the muscle is stretched by hanging a weight from its lower tendon. You can click the (+) and (−) buttons next to the Platform Height window to change the position of the platform on which the weight rests. Click on the weight again to automatically return it to the weight cabinet. The electrical stimulator displays the initial velocity of muscle shortening in the velocity window to the right of the Voltage control.

Activity 9:
Investigating the Effect of Load on Skeletal Muscle

1. Set the voltage to the maximal stimulus (8.2 volts).
2. Drag-and-drop the .5-g weight onto the muscle's lower tendon.
3. Platform height should be 75 mm.
4. Click **Stimulate** and simultaneously watch the muscle action and the oscilloscope tracing.
5. Click the **Record Data** button to retain and display the data in the grid.

What do you see happening to the muscle during the flat part of the tracing? Click **Stimulate** to repeat if you wish to see the muscle action again.

____

____

Does the force the muscle produces change during the flat part of the tracing (increase, decrease, or stay the same)?

____

Describe the muscle activity during the flat part of the tracing in terms of isotonic contraction and relaxation.

____

____

____

____

____

Circle the correct terms in parentheses in the following sequence:

The force rises during the first part of the muscle tracing due to (isometric, isotonic) contraction. The fall in force on the right side of the tracing corresponds to (isometric, isotonic) relaxation.

6. Return the .5-g weight to the cabinet. Drag the 1.5-g weight to the muscle. Click **Stimulate** and then click **Record Data.**

Which of the two weights used so far results in the highest initial velocity of shortening?

____

7. Repeat step 6 for the remaining two weights.
8. Choose **Plot Data** from the **Tools** menu.
9. Set Weight as the X-axis and Total Force as the Y-axis by dragging the slider bars.

What does the plot reveal about the resistance and the initial velocity of shortening?

____

____

____

10. Close the plot window. Click **Tools** → **Print Data** to print your data.
11. Click **Clear Table** in the data control unit at the bottom of the screen. Click **Yes** when you are asked if you want to erase all data in the table.
12. Return the current weight to the weight cabinet.

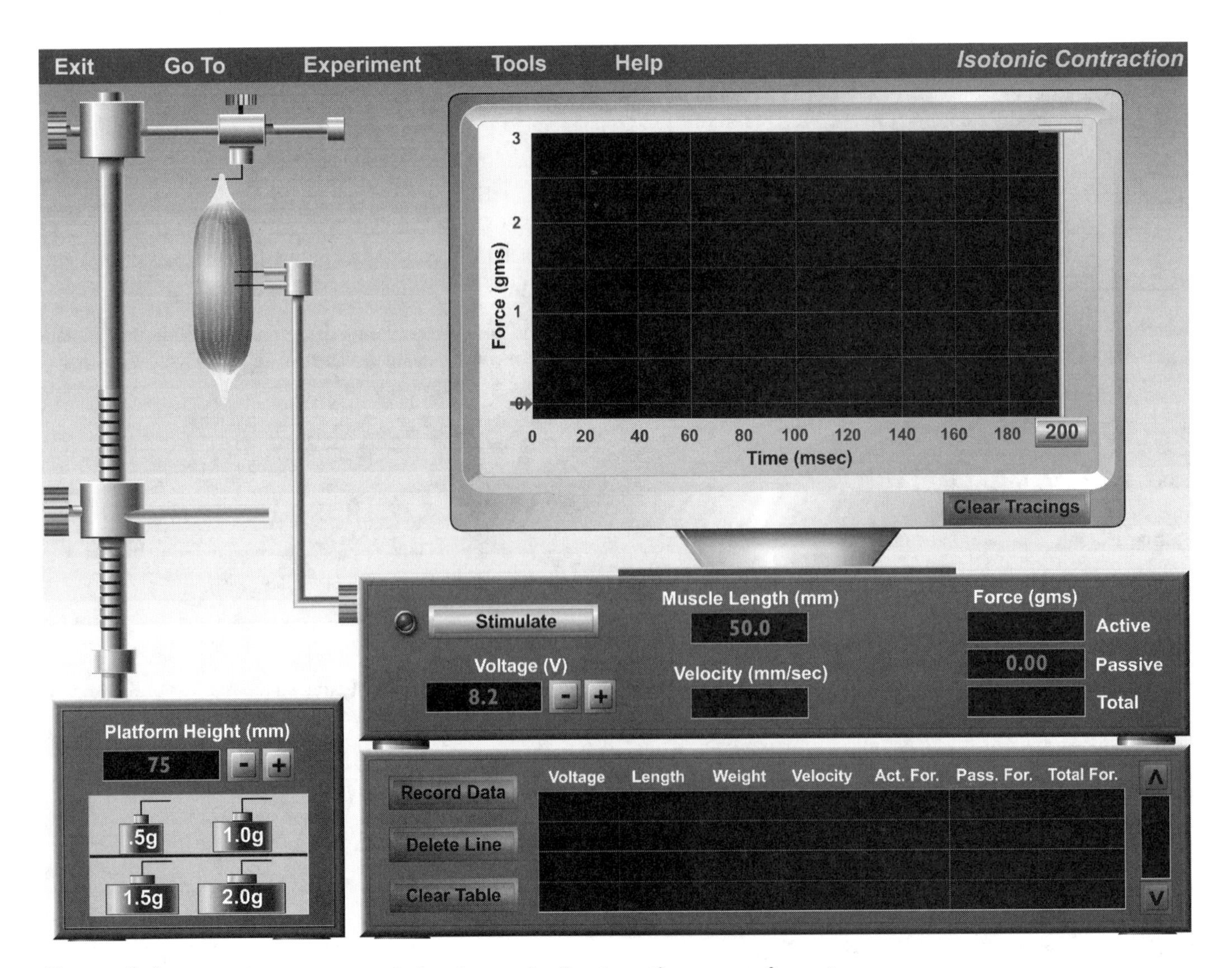

Figure 2.4 Opening screen of the Isotonic Contraction experiment.

13. Attach the 1.5-g weight to the muscle and run through the range of starting lengths from 60–90 mm in 5-mm increments. Be sure to click **Record Data** after each stimulus.

14. After all runs have been completed, choose **Plot Data** from the **Tools** menu.

15. Set Length as the X-axis and Velocity as the Y-axis by dragging the slider bars. Click **Print Plot** to print the graph.

Describe the relationship between starting length and initial velocity of shortening.

____

____

____

____

Do these results support your ideas in the isometric contraction part of this exercise?

____

16. Close the plot window. Click **Tools** → **Print Data**.

Can you set up a contraction that is completely isometric?

_______ One that is completely isotonic? _______

Explain your answers.

____

____

____

____ ■

Histology Review Supplement

Turn to p. P-142 for a review of skeletal muscle tissue.

Exercise

3 Neurophysiology of Nerve Impulses: Computer Simulation

Objectives

1. To define the following:
   irritability, conductivity, resting membrane potential, polarized, sodium-potassium pump, threshold stimulus, depolarization, action potential, repolarization, hyperpolarization, absolute refractory period, relative refractory period, nerve impulse, compound nerve action potential, and *conduction velocity*
2. To list at least four different stimuli capable of generating an action potential.
3. To list at least two agents capable of inhibiting an action potential.
4. To describe the relationship between nerve size and conduction velocity.
5. To describe the relationship between nerve myelination and conduction velocity.

The Nerve Impulse

Neurons have two major physiological properties: **irritability,** or the ability to respond to stimuli and convert them into nerve impulses, and **conductivity,** the ability to transmit an impulse (in this case, to take the neural impulse and pass it along the cell membrane). In the resting neuron (i.e., a neuron that does not have any neural impulses), the exterior of the cell membrane is positively charged and the interior of the neuron is negatively charged. This difference in electrical charge across the plasma membrane is referred to as the **resting membrane potential** and the membrane is said to be **polarized.** The **sodium-potassium pump** in the membrane maintains the difference in electrical charge established by diffusion of ions. This active transport mechanism moves 3 sodium ions out of the cell while moving in 2 potassium ions. Therefore, the major cation (positively charged ion) outside the cell in the extracellular fluid is sodium, while the major cation inside the cell is potassium. The inner surface of the cell membrane is more negative than the outer surface, mainly due to intracellular proteins, which, at body pH, tend to be negatively charged.

The resting membrane potential can be measured with a voltmeter by putting a recording electrode just inside the cell membrane with a reference, or ground, electrode outside the membrane (see Figure 3.1). In the giant squid axon (where most early neural research was conducted), or in the frog axon that will be used in this exercise, the resting membrane potential is measured at −70 millivolts (mV). (In humans, the resting membrane potential typically measures between −40 mV and −90mV.)

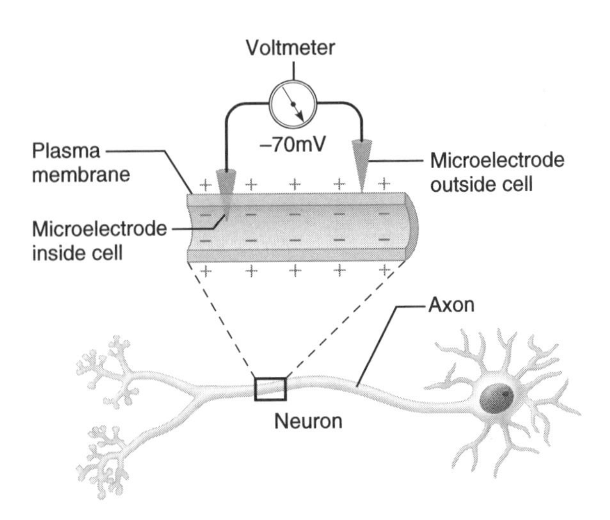

Figure 3.1

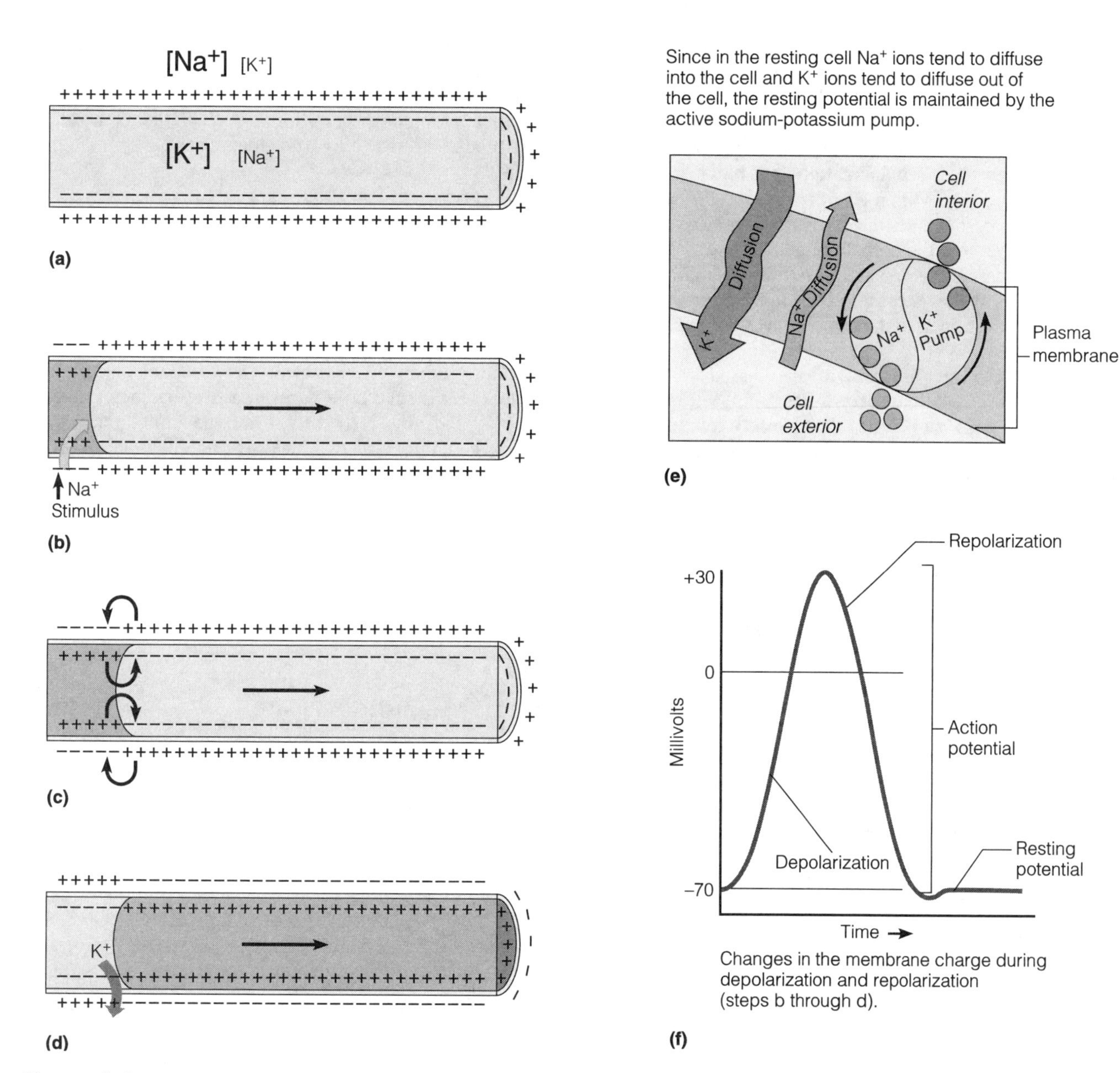

Figure 3.2 The nerve impulse. (a) Resting membrane potential (~85 mV). There is an excess of positive ions outside the cell, with Na^+ the predominant extracellular fluid ion and K^+ the predominant intracellular ion. The plasma membrane has a low permeability to Na^+. **(b)** Depolarization—reversal of the resting potential. Application of a stimulus changes the membrane permeability, and Na^+ ions are allowed to diffuse rapidly into the cell. **(c)** Generation of the action potential or nerve impulse. If the stimulus is of adequate intensity, the depolarization wave spreads rapidly along the entire length of the membrane. **(d)** Repolarization—reestablishment of the resting potential. The negative charge on the internal plasma membrane surface and the positive charge on its external surface are reestablished by diffusion of K^+ ions out of the cell, proceeding in the same direction as in depolarization. **(e)** The original ionic concentrations of the resting state are restored by the sodium-potassium pump. **(f)** A tracing of an action potential.

When a neuron is activated by a stimulus of adequate intensity, known as a **threshold stimulus,** the membrane at its *trigger zone,* typically the axon hillock, briefly becomes more permeable to sodium ions (sodium gates in the cell membrane open). Sodium ions rush into the cell, increasing the number of positive ions inside the cell and changing the membrane polarity. The interior surface of the membrane becomes less negative and the exterior surface becomes less positive, a phenomenon called **depolarization** (see Figure 3.2b). When depolarization reaches a certain point called **threshold,** an **action potential** is initiated (see Figure 3.2c) and the polarity of the membrane reverses.

When the membrane depolarizes, the resting membrane potential of −70 mV becomes less negative. When the membrane potential reaches 0 mV, indicating there is no charge difference across the membrane, the sodium ion channels start to close and potassium ion channels open. By the time the sodium ions channels finally close, the membrane potential has reached +35 mV. The opening of the potassium ion channels allows potassium ions to flow out of the cell down their electrochemical gradient—remember, like ions are repelled from each other. The flow of potassium ions out of the cell causes the membrane potential to move in a negative direction. This is referred to as **repolarization** (see Figure 3.2d). This repolarization occurs within a millisecond of the initial sodium influx and reestablishes the resting membrane potential. Actually, by the time the potassium gates close, the cell membrane has undergone a **hyperpolarization,** slipping to perhaps −75 mV. With the gates closed, the resting membrane potential is quickly returned to the normal resting membrane potential.

When the sodium gates are open, the membrane is totally insensitive to additional stimuli, regardless of the force of stimulus. The cell is in what is called the **absolute refractory period.** During repolarization, the membrane may be stimulated if a very strong stimulus is used. This period is called the **relative refractory period.**

The action potential, once started, is a self-propagating phenomenon, spreading rapidly along the neuron membrane. The action potential follows the *all-or-none* law, in which the neuron membrane either depolarizes 100% or not at all. In neurons, the action potential is also called a **nerve impulse.** When it reaches the axon terminal, it triggers the release of neurotransmitters into the synaptic cleft. Depending on the situation, the neurotransmitter will either excite or inhibit the postsynaptic neuron.

In order to study nerve physiology, we will use a frog nerve and several electronic instruments. The first instrument is the *electronic stimulator.* Nerves can be stimulated by chemicals, touch, or electric shock. The electronic stimulator administers an electric shock that is pure direct current (DC), and allows duration, frequency, and voltage of the shock to be precisely controlled. The stimulator has two output terminals; the positive terminal is red and the negative terminal is black. Voltage leaves the stimulator via the red terminal, passes through the item to be stimulated (in this case, the nerve), and returns to the stimulator at the black terminal to complete the circuit.

The second instrument is the **oscilloscope,** an instrument that measures voltage changes over a period of time. The face of the oscilloscope is similar to a black-and-white TV screen. The screen of the oscilloscope is the front of a tube with a filament at the other end. The filament is heated and gives off a beam of electrons. The beam passes to the front of the tube. Electronic circuitry allows for the electron beam to be brought across the screen in preset time intervals. When the electrons hit the phosphorescent material on the inside of the screen, a spot on the screen will glow. When we apply a stimulus to a nerve, the oscilloscope screen will display one of the following three results: no response, a flat line, or a graph with a peak. A graph with a peak indicates that an action potential has been generated.

While performing the following experiments, keep in mind that you are working with a nerve, which consists of many neurons—you are not working with just a single neuron. The action potential you will see on the oscilloscope screen reflects the cumulative action potentials of all the neurons in the nerve, called a **compound nerve action potential.** Although an action potential follows the all-or-none law within a single neuron, it does not necessarily follow this law within an entire nerve. When you electrically stimulate a nerve at a given voltage, the stimulus may result in the depolarization of most of the neurons, but not necessarily all of them. To achieve depolarization of *all* of the neurons, a higher stimulus voltage may be needed.

Eliciting a Nerve Impulse

In the following experiments, you will be investigating what kinds of stimuli trigger an action potential. To begin, select **Neurophysiology of Nerve Impulses** from the main menu. The opening screen will appear in a few seconds (see Figure 3.3). Note that a sciatic nerve from a frog has been placed into the nerve chamber. Leads go from the stimulator output to the nerve chamber, the vertical box on the left side. Leads also go from the nerve chamber to the oscilloscope. Notice that these leads are red and black. The stimulus travels along the red lead to the nerve. When the nerve depolarizes, it will generate an electrical impulse that will travel along the red wire to the oscilloscope and back to the nerve along the black wire.

Activity 1: Electrical Stimulation

1. Set the voltage at 1.0 V by clicking the (+) button next to the **Voltage** display.
2. Click **Single Stimulus.**

Do you see any kind of response on the oscilloscope screen?

____

If you saw no response, or a flat line indicating no action potential, click the **Clear** button on the oscilloscope, increase the voltage, and click **Single Stimulus** again until you see a trace (deflection of the line) that indicates an action potential.

What was the *threshold voltage,* that is, the voltage at which you first saw an action potential?

________ V

Click **Record Data** on the data collection box to record your results.

3. If you wish to print your graph, click **Tools** and then **Print Graph.** You may do this each time you generate a graph on the oscilloscope screen.
4. Increase the voltage by 0.5 V and click **Single Stimulus.**

How does this tracing compare to the one trace that was generated at the threshold voltage? (*Hint:* look very carefully at the tracings.)

____

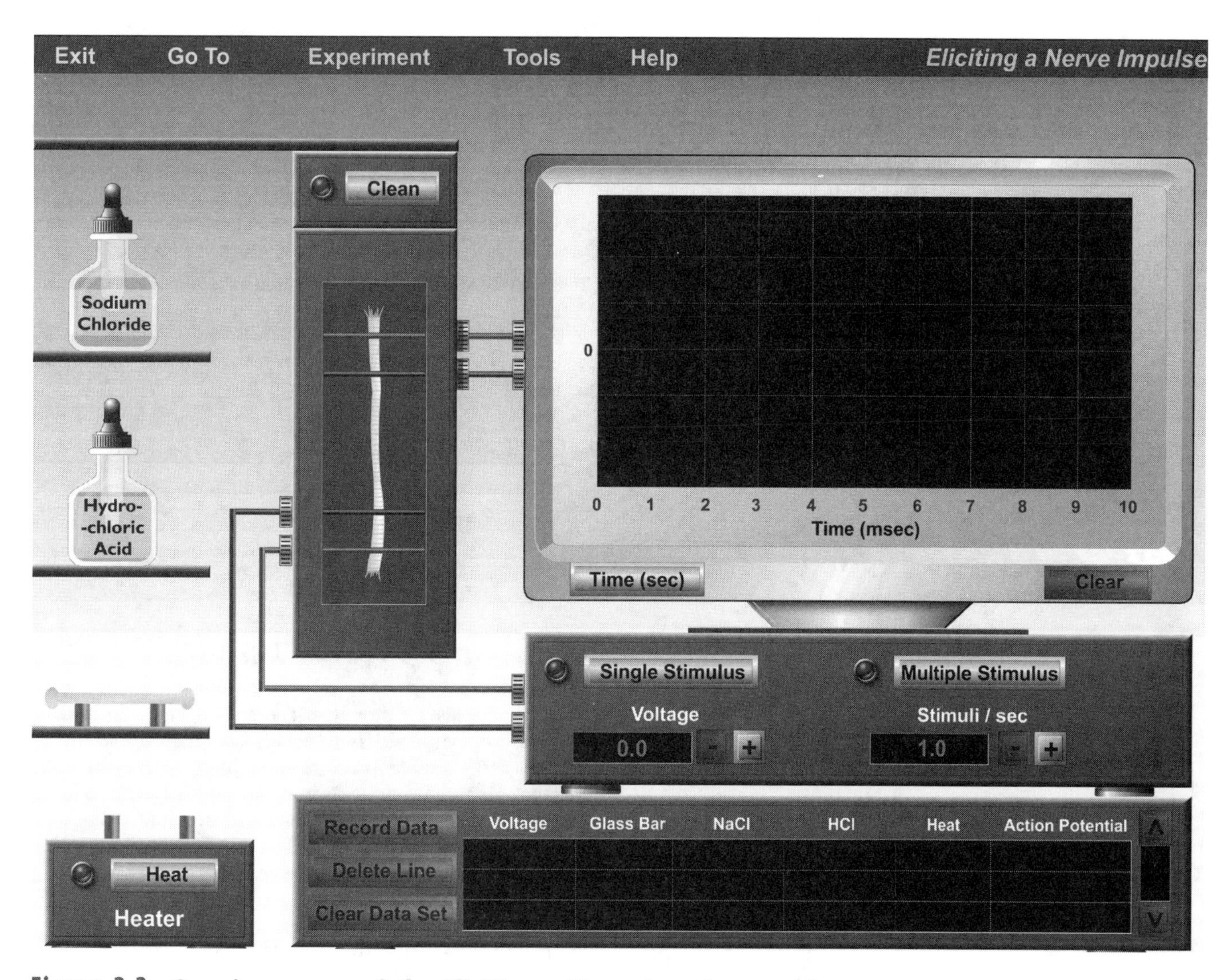

Figure 3.3 Opening screen of the Eliciting a Nerve Impulse experiment.

What reason can you give for your answer?

____

____

____

Click **Record Data** on the data collection box to record your results.

5. Continue to increase the voltage by 0.5 V and to click **Single Stimulus** until you find the point beyond which no further increase occurs in the peak of the action potential trace.

Record this maximal voltage here: _______ V

Click **Record Data** to record your results. ■

Now that we have seen that an electrical impulse can cause an action potential, let's try some other methods of stimulating a nerve.

Activity 2: Mechanical Stimulation

1. Click the **Clear** button on the oscilloscope.

2. Using the mouse, click the glass rod located on the bottom shelf an the left side of the screen, and drag it over to the nerve. When the glass rod is over the nerve, release the mouse button to indicate that the rod is now touching the nerve. What do you see on the oscilloscope screen?

____

How does this tracing compare with the other tracings that you have generated?

____

____

Click **Record Data** to record your results. Leave the graph on the screen so that you can compare it to the graph you will generate in the next activity. ■

Activity 3: Thermal Stimulation

1. Click on the glass rod and drag it to the heater, releasing the mouse button. Click on the **Heat** button. When the rod turns red, indicating that it has been heated, click and drag the rod over the nerve and release the mouse button. What happens?

____

____

How does this trace compare to the trace that was generated with the unheated glass rod?

____

____

What explanation can you provide for this?

____

____

Click **Record Data** to record your results. Then click **Clear** to clear the oscilloscope screen for the next activity. ■

Activity 4: Chemical Stimulation

1. Click and drag the dropper from the bottle of sodium chloride (salt solution) over to the nerve in the chamber and then release the mouse button to dispense drops.

Does this generate an action potential? ______________

2. Using your threshold setting, stimulate the nerve.

Does this tracing differ from the original threshold stimulus tracing?

____

Click **Record Data** to record your results.

3. Click the **Clean** button on top of the nerve chamber. This will return the nerve to its original (nonsalted) state. Click **Clear** to clear the oscilloscope screen.

4. Click and drag the dropper from the bottle of hydrochloric acid over to the nerve, and release the mouse button to dispense drops.

Does this generate an action potential? ______________

5. Does this tracing differ from the one generated by the original threshold stimulus?

____

Click **Record Data** to record your results.

6. Click on the **Clean** button on the nerve chamber to clean the chamber and return the nerve to its untouched state.

Click **Tools** → **Print Data** to print the data you have recorded for this experiment.

To summarize your experimental results, what kinds of stimuli can elicit an action potential?

____

____

______________________________ ■

Inhibiting a Nerve Impulse

Numerous physical factors and chemical agents can impair the ability of nerve fibers to function. For example, deep pressure and cold temperature both block nerve impulse transmission by preventing local blood supply from reaching the nerve fibers. Local anesthetics, alcohol, and numerous other chemicals are also effective in blocking nerve transmission. In this experiment, we will study the effects of various agents on nerve transmission.

To begin, click the **Experiment** pull-down menu and select **Inhibiting a Nerve Impulse.** The display screen for this activity is similar to the screen in the first activity (see Figure 3.4). To the left are bottles of several agents that we will test on the nerve. Keep the tracings you printed out from the first activity close at hand for comparison.

Activity 5: Testing the Effects of Ether

1. Using the mouse, click and drag the dropper from the bottle marked *ether* over to the nerve, in between the stimulating electrodes and recording electrodes. Release the mouse button to dispense drops.

2. Click **Stimulate,** using the voltage setting from the threshold stimulus you used in the earlier activities. What sort of trace do you see?

____

What has happened to the nerve? ______________

____

Click **Record Data** to record your results.

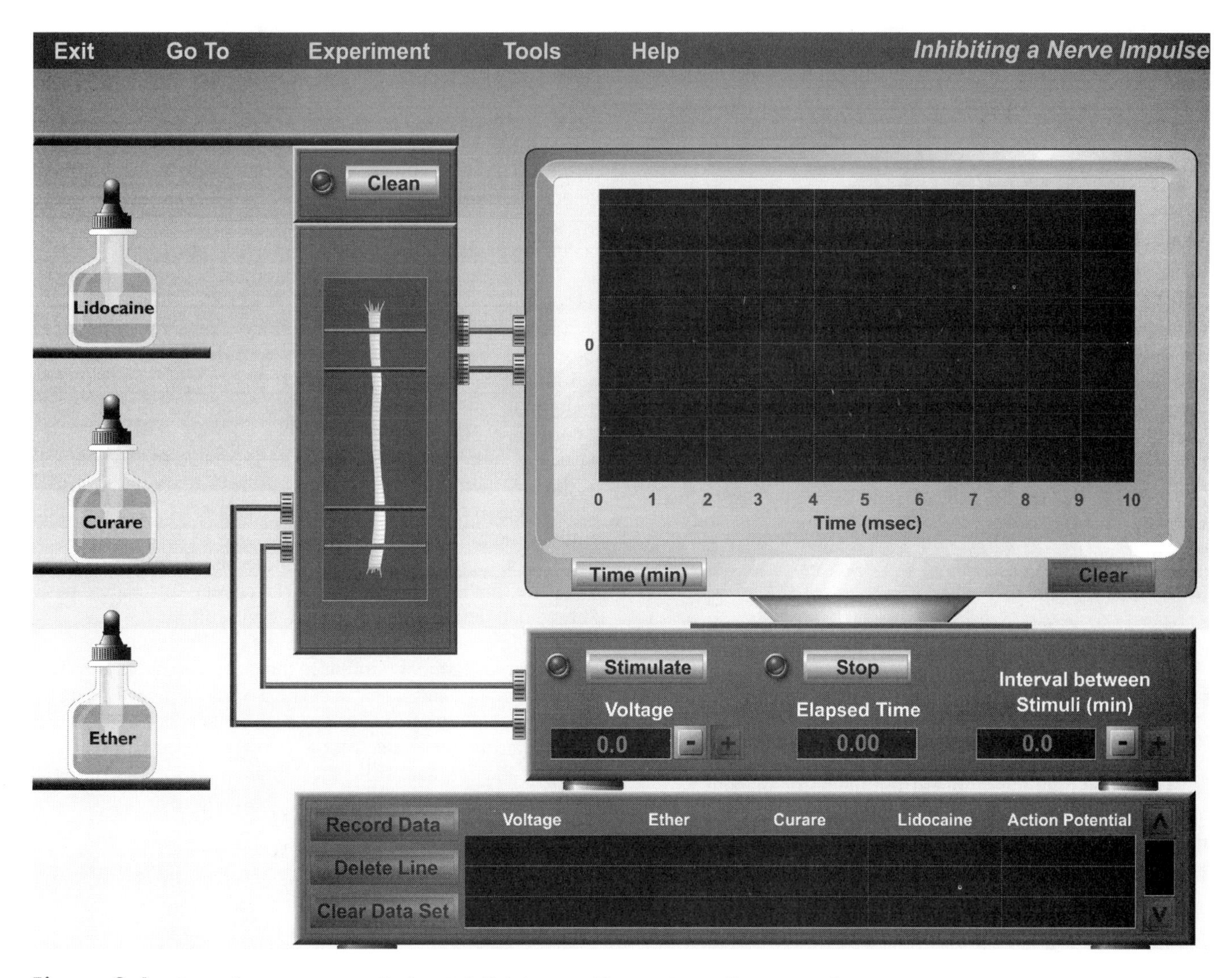

Figure 3.4 Opening screen of the Inhibiting a Nerve Impulse experiment.

3. Click on the **Time (min.)** button on the oscilloscope. The screen will now display activity over the course of 10 minutes (the space between each vertical line representing 1 minute). Because of the change in time scale, an action potential will look like a sharp vertical spike on the screen.

4. Click the (+) button under **Interval between Stimuli** on the stimulator until the timer is set for 2.0 minutes. This will set the stimulus to stimulate the nerve every two minutes. Click on **Stimulate** to start the stimulations. Watch the **Elapsed Time** display.

How long does it take for the nerve to return to normal?

____

5. Click on the **Stop** button to stop this action and to return the Elapsed Time to 0.0.

6. Click the **Time (msec)** button on the oscilloscope to return it to its normal millisecond display.

7. Click **Clear** to clear the oscilloscope for the next activity.

8. Click the (−) button under **Interval between Stimuli** until it is reset to 0.00. ■

Activity 6:
Testing the Effects of Curare

Curare is a well-known plant extract that South American Indians used to paralyze their prey. It is an alpha-toxin that binds to acetylcholine binding sites on the postsynaptic cell membrane, which will prevent the acetylcholine from acting. Curare blocks synaptic transmission by preventing the flow of neural impulses from neuron to neuron.

1. Click and drag the dropper from the bottle marked *curare* and position the dropper on the nerve, in between the stimulating and recording electrodes. Release the mouse button to dispense drops.

2. Set the stimulator at the threshold voltage and stimulate the nerve. What effect on the action potential is noted?

____

What explains this effect? ______________________

What do you think would be the overall effect of curare on the organism?

____

____

Click **Record Data** to record your results.

3. Click on the **Clean** button on the nerve chamber to remove the curare and return the nerve to its original untouched state.

4. Click **Clear** to clear the oscilloscope screen for the next activity. ■

Activity 7:
Testing the Effects of Lidocaine

Note: lidocaine is a sodium-channel antagonist.

1. Click and drag the dropper from the bottle marked *lidocaine* and position it over the nerve, between the stimulating and recording electrodes. Release the mouse button to dispense drops. Does this generate a trace?

____

2. Stimulate the nerve at the threshold voltage. What sort of tracing is seen?

____

Why does lidocaine have this effect on nerve fiber transmission?

____

____

Click **Record Data** to record your results. Click **Tools** → **Print Data** to print your data.

3. Click on the **Clean** button on the nerve chamber to remove the lidocaine and return the nerve to its original untouched state. ■

Nerve Conduction Velocity

As has been pointed out, one of the major physiological properties of neurons is conductivity: the ability to transmit the nerve impulse to other neurons, muscles, or glands. The nerve impulse, or propagated action potential, occurs when sodium ions flood into the neuron, causing the membrane to depolarize. Although this event is spoken of in electrical terms, and is measured using instruments that measure electrical events, the velocity of the action potential along a neural membrane does not occur at the speed of light. Rather, this event is much slower. In certain nerves in the human, the velocity of an action potential may be as fast as 120 meters per second. In other nerves, conduction speed is much slower, occurring at a speed of less than 3 meters per second.

To see the setup for this experiment, click the **Experiment** pull-down menu and select **Nerve Conduction Velocity** (Figure 3.5). In this exercise, the oscilloscope and stimulator will be used along with a third instrument, the bio-amplifier. The **bio-amplifier** is used to amplify any membrane depolarization so that the oscilloscope can easily record the event. Normally, when a membrane depolarization sufficient to initiate action potential is looked at, the interior of the cell membrane goes from -70 mV to about $+40$ mV. This is easily registered and viewable on an oscilloscope, without the aid of an amplifier. However, in this experiment, it is the change in the membrane potential on the *outside* of the nerve that is being observed. The change that occurs here during depolarization will be so minuscule that it must be amplified in order to be visible on the oscilloscope.

A nerve chamber (similar to the one used in the previous two experiments) will be used. The design is basically a plastic box with platinum electrodes running across it. The nerve will be laid on these electrodes. Two electrodes will be used to bring the impulse from the stimulator to the nerve and three will be used for recording the membrane depolarization.

In this experiment, we will determine and compare the conduction velocities of different types of nerves. We will examine four nerves: an earthworm nerve, a frog nerve, and two rat nerves. The earthworm nerve is the smallest of the four. The frog nerve is a medium-sized myelinated nerve. Rat nerve 1 is a medium-sized unmyelinated nerve. Rat nerve 2 is a large, myelinated nerve—the largest nerve in this group. We will observe the effects of size and myelination on nerve conductivity.

The basic layout of the materials is shown in Figure 3.5. The two wires (red and black) from the stimulator connect with the top right side of the nerve chamber. Three wires (red, black, and a bare wire cable) are attached to connectors on the other end of the nerve chamber and go to the bio-amplifier. The bare cable serves as a "ground reference" for the electrical circuit and provides the reference for comparison of any change in membrane potential. The bio-amplifier is connected to the oscilloscope so that any amplified membrane changes can be observed. The stimulator output, called the *pulse*, has been connected to the oscilloscope so that when the nerve is stimulated, the tracing will start across the oscilloscope screen. Thus, the time from the start of the trace on the left-hand side of the screen (when the nerve was stimulated) to the actual nerve deflection (from the recording electrodes) can be accurately measured. This amount of time, usually in milliseconds, is critical for determining conduction velocity.

Look closely at the screen. The wiring of the circuit may seem complicated but actually is not. First, look at the stimulator, found on top of the oscilloscope. On the left side, red and black wires leave the stimulator to go to the nerve cham-

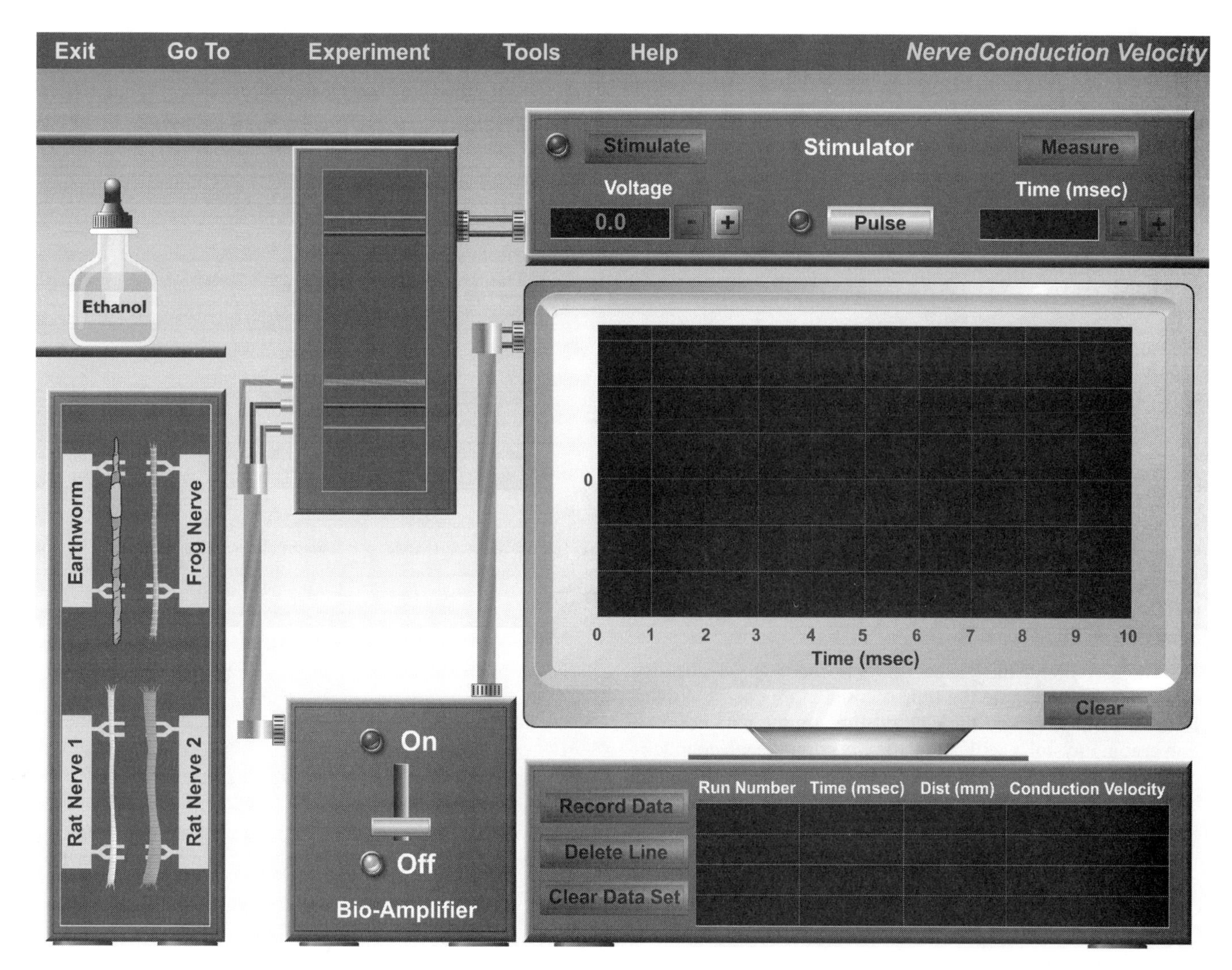

Figure 3.5 Opening screen of the Nerve Conduction Velocity experiment.

ber. Remember, the red wire is the *hot* wire that carries the impulse from the stimulator and the black wire is the return to the stimulator that completes the circuit. When the nerve is stimulated, the red recording wire (leaving the left side of the nerve chamber) will pick up the membrane impulse and bring it to the bio-amplifier. The black wire, as before, completes the circuit, and the bare cable wire simply acts as a reference electrode. The membrane potential, picked up by the red wire, is then amplified by the bio-amplifier and the output is carried to the oscilloscope. The oscilloscope then shows the trace of the nerve action potential.

Activity 8: Measuring Nerve Conduction Velocity

1. On the stimulator, click the **Pulse** button.

2. Turn the bio-amplifier on by clicking the horizontal bar on the bio-amplifier and dragging it to the **On** setting.

On the left side of the screen are the four nerves that will be studied. The nerves included are the earthworm, a frog nerve, and two rat nerves of different sizes. The earthworm as a whole is used because it has a nerve running down its ventral surface. A frog nerve is used as the frog has long been the animal of choice in many physiology laboratories. The rat nerves are used so that you may compare (a) the conduction velocity of different sized nerves and (b) the conduction velocity of a myelinated versus unmyelinated nerve. Remember that the frog nerve is myelinated and that rat nerve 1 is the same size as the frog nerve but unmyelinated. Rat nerve 2, the largest nerve of the bunch, is myelinated.

3. Using the mouse, click and drag the dropper from the bottle labeled *ethanol* over the earthworm and release the mouse button to dispense drops of ethanol. This will narcotize the worm so it does not move around during the experiment but will not affect nerve conduction velocity. The alcohol is at a low enough percentage that the worm will be fine and back to normal within 15 minutes.

4. Click and drag the earthworm into the nerve chamber. Be sure the worm is over both of the stimulating electrodes and all three of the recording electrodes.

5. Using the (+) button next to the **Voltage** display, set the voltage to 1.0 V. Then click **Stimulate** to stimulate the nerve. Do you see an action potential? If not, increase the voltage by increments of 1.0 V until a trace is obtained.

At what threshold voltage do you first see an action potential generated?

_______ V

6. Next, click on the **Measure** button located on the stimulator. You will see a vertical yellow line appear on the far left edge of the oscilloscope screen. Now click the (+) button under the Measure button. This will move the yellow line to the right. This line lets you measure how much time has elapsed on the graph at the point that the line is crossing the graph. You will see the elapsed time appear on the **Time (msec)** display on the stimulator. Keep clicking (+) until the yellow line is right at the point in the graph where the graph ceases being a flat line and first starts to rise.

7. Once you have the yellow line positioned at the start of the graph's ascent, note the time elapsed at this point. Click **Record Data** to record the elapsed time on the data collection graph. PhysioEx will automatically compute the conduction velocity based on this data. Note that the data collection box includes a **Distance (mm)** column and that the distance is always 43 mm. This is the distance from the red stimulating wire to the red recording wire. In a wet lab, you would have to measure the distance yourself before you could proceed with calculating the conduction velocity.

It is important that you have the yellow vertical measuring line positioned at the start of the graph's rise before you click **Record Data**—otherwise, the conduction velocity calculated for the nerve will be inaccurate.

8. Fill in the data under the Earthworm column on the chart below:

9. Click and drag the earthworm to its original place. Click **Clear** to clear the oscilloscope screen.

10. Repeat steps 4 through 9 for the remaining nerves. Remember to click **Record Data** after each experimental run and to fill in the chart for question 8.

11. Click **Tools** → **Print Data** to print your data.

Which nerve in the group has the slowest conduction velocity?

____

What was the speed of the nerve? _______________

Which nerve in the group of four has the fastest conduction velocity?

____

What was the speed of the nerve? _______________

What is the relationship between nerve size and conduction velocity? _______________________________________

____

____

____

Based on the results, what is your conclusion regarding conduction velocity and whether the nerve is myelinated or not?

____

____

____

____

Nerve	**Earthworm (small nerve)**	**Frog (medium nerve, myelinated)**	**Rat nerve 1 (medium nerve, unmyelinated)**	**Rat nerve 2 (large nerve, myelinated)**
Threshold voltage				
Elapsed time from stimulation to action potential				
Conduction velocity				

What is the major reason for the differences seen in conduction velocity between the myelinated nerves and the unmyelinated nerves? ______________________________

____

____

____ ■

History Review Supplement

Turn to p. P-143 for a review of nervous tissue.

Exercise 4 Endocrine System Physiology: Computer Simulation

Objectives

1. To define the following:
   metabolism, hormone replacement therapy, diabetes type I, diabetes type II, and *glucose standard curve.*
2. To explain the role of thyroxine in maintaining an animal's metabolic rate.
3. To explain the effects of thyroid-stimulating hormone on an animal's metabolic rate.
4. To understand how estrogen affects uterine tissue growth.
5. To explain how hormone replacement therapy works.
6. To explain why insulin is important, and how it can be used to treat diabetes.

The endocrine system exerts many complex and interrelated effects on the body as a whole, as well as on specific tissues and organs. Studying the effects of hormones on the body is difficult to do in a wet lab because experiments often take days, weeks, or even months to complete and are expensive. In addition, live animals may need to be sacrificed, and technically difficult surgical procedures are sometimes necessary. This computer simulation allows you to study the effects of given hormones on the body by using "virtual" animals rather than live ones. You can carry out delicate surgical techniques with the click of a button, and complete experiments in a fraction of the time that it would take in an actual wet lab environment.

Hormones and Metabolism

Metabolism is the broad term used for all biochemical reactions occurring in the body. Metabolism involves *catabolism,* a process by which complex materials are broken down into simpler substances, usually with the aid of enzymes found in the body cells. Metabolism also involves *anabolism,* in which the smaller materials are built up by enzymes to build larger, more complex molecules. When larger molecules are made, energy is stored in the various bonds formed. When bonds are broken in catabolism, energy that was stored in the bonds is released for use by the cell. Some of the energy liberated may go into the formation of ATP, the energy-rich material used by the body to run itself. However, not all of the energy liberated goes into this pathway; some is given off as body heat. Humans are *homeothermic* animals, meaning they have a fixed body temperature. Maintaining this temperature is important to maintaining the metabolic pathways found in the body.

The most important hormone in maintaining metabolism and body heat is *thyroxine,* the hormone of the thyroid gland, found in the neck. The thyroid gland secretes thyroxine, but the production of thyroxine is really controlled by the pituitary gland, which secretes *thyroid-stimulating hormone (TSH).* TSH is carried by the blood to the thyroid gland (its *target tissue*) and causes the thyroid to produce more thyroxine. So in an indirect way, an animal's metabolic rate is the result of pituitary hormones.

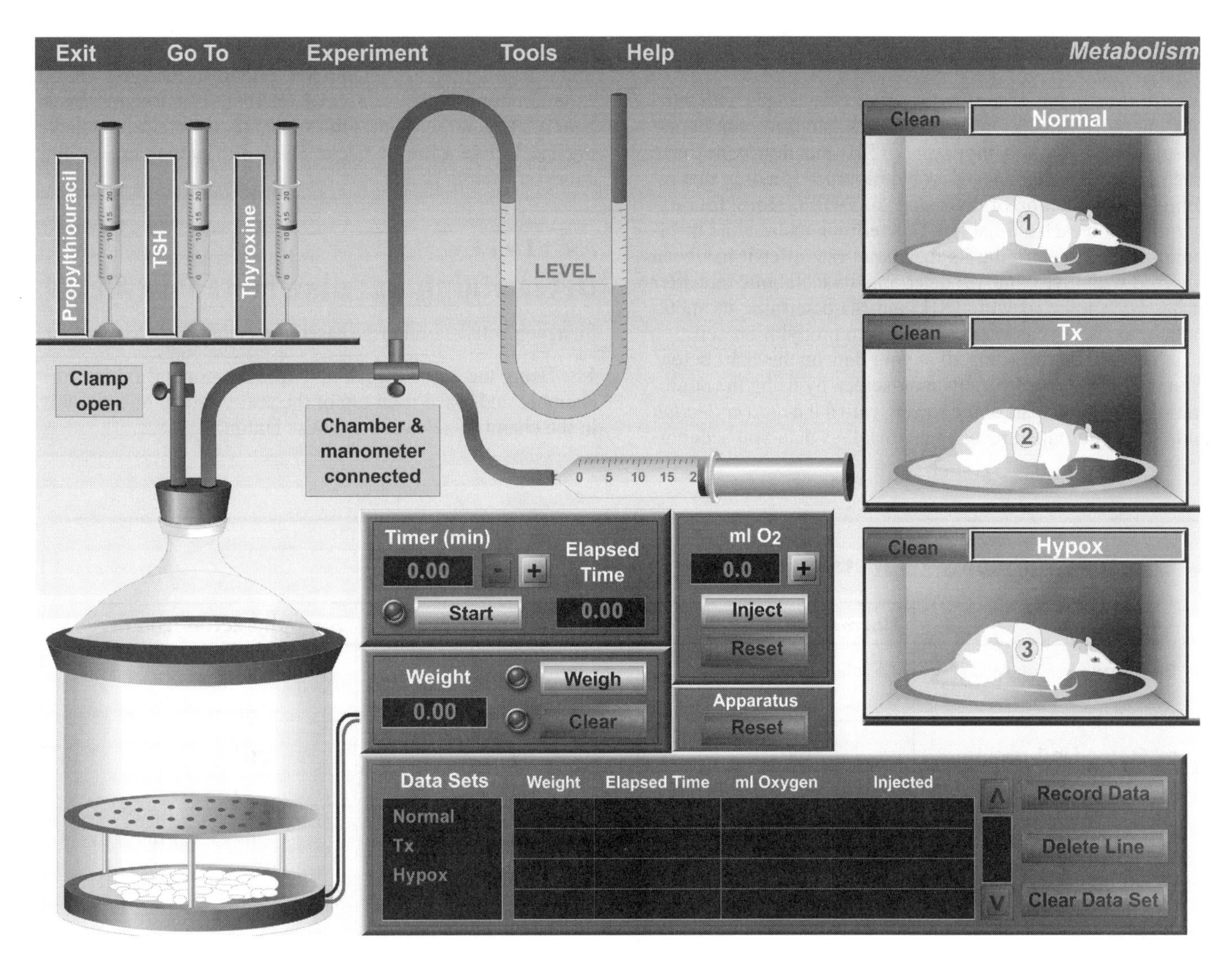

Figure 4.1 Opening screen of the Metabolism experiment.

In the following experiments, you will investigate the effects of thyroxine and TSH on an animal's metabolic rate. To begin, select **Endocrine System Physiology** from the main menu. The opening screen will appear in a few seconds (see Figure 4.1). Select **Balloons On** from the Help menu for help identifying the equipment on-screen (you will see labels appear as you roll over each piece of equipment). Select **Balloons Off** to turn this feature off before you begin the experiments.

Study the screen. You will see a jar-shaped chamber to the left, connected to a *respirometer-manometer apparatus* (consisting of a U-shaped tube, a syringe, and associated tubing.) You will be placing animals—in this case, rats—in the chamber in order to gather information about how thyroxine and TSH affect their metabolic rates. Note that the chamber also includes a weight scale, and next to the chamber is a timer for setting and timing the length of a given experiment. Under the timer is a weight display.

Two tubes are connected to the top of the chamber. The left tube has a clamp on it that can be opened or closed. Leaving the clamp open allows outside air into the chamber; closing the clamp creates a closed, airtight system. The other tube leads to a *T-connector*. One branch of the T leads to a fluid-containing U-shaped tube, called a *manometer*. As an animal uses up the air in the closed system, this fluid will rise in the left side of the U-shaped tube and fall in the right.

The other branch of the T-connector leads to a syringe filled with air. Using the syringe to inject air into the tube, you will measure the amount of air that is needed to return the fluid columns to their original levels. This measurement will be equal to the amount of oxygen used by the animal during the elapsed time of the experiment. Soda lime, found at the bottom of the chamber, absorbs the carbon dioxide given off by the animal so that the amount of oxygen used can easily be measured. The amount of oxygen used by the animal, along with its weight, will be used to calculate the animal's metabolic rate.

Also on the screen are three white rats in their individual cages. These are the specimens you will use in the following experiments. One rat is *normal;* the second is *thyroidectomized* (abbreviated on the screen as *Tx*)— meaning its thyroid has been removed; and the third is *hypophysectomized* (abbreviated on the screen as *Hypox*), meaning its pituitary gland has been removed. The pituitary gland is also known as the *hypophysis,* and removal of this organ is called a *hypophysectomy.*

To the top left of the screen are three syringes with various chemicals inside: propylthiouracil, thyroid-stimulating hormone (TSH), and thyroxine. TSH and thyroxine have been previously mentioned; propylthiouracil is a drug that inhibits the production of thyroxine. You will perform four experiments on each animal to: (1) determine its baseline metabolic rate, (2) determine its metabolic rate after it has been injected with thyroxine, (3) determine its metabolic rate after it has been injected with TSH, and (4) determine its metabolic rate after it has been injected with propylthiouracil.

You will be recording all of your data on the chart below. You may also record your data on-screen by using the equipment in the lower part of the screen, called the *data collection unit.* This equipment records and displays data you accumulate during the experiments. Check that the data set for **Normal** is highlighted in the **Data Sets** window, since you will be experimenting with the normal rat first. The **Record Data** button lets you record data after an experimental trial. Clicking the **Delete Line** or **Clear Data Set** buttons erases any data you want to delete.

Activity 1: Determining Baseline Metabolic Rates

First, you will determine the baseline metabolic rate for each rat.

1. Using the mouse, click and drag the **normal** rat into the chamber and place it on top of the scale. When the animal is in the chamber, release the mouse button.

Effects of Hormones on Metabolic Rate

	Normal rat	Thyroidectomized rat	Hypophysectomized rat
Baseline			
Weight	______ grams	______ grams	______ grams
ml O_2 used in 1 minute	______ ml	______ ml	______ ml
ml O_2 used per hour	______ ml	______ ml	______ ml
Metabolic rate	______ ml O_2/kg/hr	______ ml O_2/kg/hr	______ ml O_2/kg/hr
With thyroxine			
Weight	______ grams	______ grams	______ grams
ml O_2 used in 1 minute	______ ml	______ ml	______ ml
ml O_2 used per hour	______ ml	______ ml	______ ml
Metabolic rate	______ ml O_2/kg/hr	______ ml O_2/kg/hr	______ ml O_2/kg/hr
With TSH			
Weight	______ grams	______ grams	______ grams
ml O_2 used in 1 minute	______ ml	______ ml	______ ml
ml O_2 used per hour	______ ml	______ ml	______ ml
Metabolic rate	______ ml O_2/kg/hr	______ ml O_2/kg/hr	______ ml O_2/kg/hr
With propylthiouracil			
Weight	______ grams	______ grams	______ grams
ml O_2 used in 1 minute	______ ml	______ ml	______ ml
ml O_2 used per hour	______ ml	______ ml	______ ml
Metabolic rate	______ ml O_2/kg/hr	______ ml O_2/kg/hr	______ ml O_2/kg/hr

2. Be sure the clamp on the left tube (on top of the chamber) is open, allowing air to enter the chamber. If the clamp is closed, click on it to open it.

3. Be sure the indicator next to the T-connector reads "Chamber and manometer connected." If not, click on the **T-connector knob.**

4. Click on the **Weigh** button in the box to the right of the chamber to weigh the rat. Record this weight in the Baseline section of the chart on p. P-38 in the row labeled "Weight."

5. Click the (+) button on the **Timer** so that the Timer display reads 1 minute.

6. Click on the clamp to close it. This will prevent any outside air from entering the chamber and ensure that the only oxygen the rat is breathing is the oxygen inside the closed system.

7. Click **Start** on the Timer display. You will see the elapsed time appear in the "Elapsed Time" display. Watch what happens to the water levels in the U-shaped tube.

8. At the end of the 1-minute period, the timer will automatically stop. When it stops, click on the **T-connector knob** so that the indicator reads "Manometer and syringe connected."

9. Click on the clamp to open it so that the rat can once again breathe outside air.

10. Click the (+) button under ml O_2 (below the syringe) so that the display reads "1.0". Then click **Inject,** and watch what happens to the fluid levels. Continue clicking the (+) button and injecting air until the fluid in the two arms of the U-tube are level again. How many milliliters of air need to be added to equalize the fluid in the two arms? (This is equivalent to the amount of oxygen that the rat used up during 1 minute in the closed chamber.) Record this measurement in the Baseline section of the chart on p. P-38 in the row labeled "ml O_2 used in 1 minute."

11. Determine the oxygen consumption per hour for the rat. Use the following formula:

$$\frac{\text{ml } O_2 \text{ consumed}}{1 \text{ minute}} \times \frac{60 \text{ minutes}}{1 \text{ hr}} = \text{ml } O_2/\text{hr}$$

Record this data in the Baseline section of the chart in the row labeled "ml O_2 used per hour."

12. Now that you have the amount of oxygen used per hour, determine the metabolic rate per kilogram of body weight by using the following formula. (Note that you will need to convert the weight data from g to kg before you can use the formula.)

$$\text{Metabolic Rate} = \frac{\text{ml } O_2/\text{hr}}{\text{wt. in kg}} = \text{\_\_\_\_\_\_\_\_\_\_ ml } O_2/\text{kg/hr}$$

Record this data in the Baseline section of the chart in the row labeled "Metabolic rate."

13. Click **Record Data.**

14. Click and drag the rat from the chamber back to its cage.

15. Click the **Reset** button in the box labeled *Apparatus*.

16. Now repeat steps 1–15 for the thyroidectomized (Tx) and hypophysectomized (Hypox) rats. Record your data in the Baseline section of the chart under the corresponding column for each rat. Be sure to highlight **Tx** under **Data Sets** (on the data collection box) before beginning the experiment on the thyroidectomized rat; likewise, highlight **Hypox** under **Data Sets** before beginning the experiment on the hypophysectomized rat.

How did the metabolic rates of the three rats differ?

____

____

____

Why did the metabolic rates differ?

____

____

____ ■

Activity 2:
Determining the Effect of Thyroxine on Metabolic Rate

Next, you will investigate the effects of thyroxine injections on the metabolic rates of all three rats.

Note that in a wet lab environment, you would normally need to inject thyroxine (or any other hormone) into a rat *daily* for a minimum of 1–2 weeks in order for any response to be seen. However, in the following simulations, you will inject the rat only once and be able to witness the same results as if you had administered multiple injections over the course of several weeks. In addition, by clicking the **Clean** button while a rat is inside its cage, you can magically remove all residue of any previously injected hormone from the rat and perform a new experiment on the same rat. In a real wet lab environment, you would need to either wait weeks for hormonal residue to leave the rat's system or use a different rat.

1. Select a rat to test. You will eventually test all three, and it doesn't matter what order you test them in. Under **Data Sets,** highlight **Normal, Tx,** or **Hypox** depending on which rat you select.

2. Click the **Reset** button in the box labeled *Apparatus.*

3. Click on the syringe labeled **thyroxine** and drag it over to the rat. Release the mouse button. This will cause thyroxine to be injected into the rat.

4. Click and drag the rat into the chamber. Perform steps 1–12 of Activity 1 again, except this time record your data in the With Thyroxine section of the chart.

5. Click **Record Data.**

6. Click and drag the rat from the chamber back to its cage and click **Clean** to cleanse it of all traces of thyroxine.

7. Now repeat steps 1–6 for the remaining rats. Record your data in the With Thyroxine section of the chart under the corresponding column for each rat.

What was the effect of thyroxine on the normal rat's metabolic rate? How does it compare to the normal rat's baseline metabolic rate?

____

____

Why was this effect seen? ______________________

____

What was the effect of thyroxine on the thyroidectomized rat's metabolic rate? How does it compare to the thyroidectomized rat's baseline metabolic rate?

____

____

____

Why was this effect seen? ______________________

____

What was the effect of thyroxine on the hypophysectomized rat's metabolic rate? How does it compare to the hypophysectomized rat's baseline metabolic rate?

____

____

____

Why was this effect seen? ______________________

____ ■

Activity 3: Determining the Effect of TSH on Metabolic Rate

Next, you will investigate the effects of TSH injections on the metabolic rates of the three rats. Select a rat to experiment on first, and then proceed.

1. Under **Data Sets,** highlight **Normal, Tx,** or **Hypox** depending on which rat you are using.
2. Click the **Reset** button in the box labeled *Apparatus.*
3. Click and drag the syringe labeled **TSH** over to the rat and release the mouse button, injecting the rat.
4. Click and drag the rat into the chamber. Perform steps 1–12 of Activity 1 again. Record your data in the With TSH section of the chart.
5. Click **Record Data.**
6. Click and drag the rat from the chamber back to its cage and click **Clean** to cleanse it of all traces of TSH.
7. Now repeat this activity for the remaining rats. Record your data in the With TSH section of the chart under the corresponding column for each rat.

What was the effect of TSH on the normal rat's metabolic rate? How does it compare to the normal rat's baseline metabolic rate?

____

____

____

Why was this effect seen? ______________________

____

What was the effect of TSH on the thyroidectomized rat's metabolic rate? How does it compare to the thyroidectomized rat's baseline metabolic rate?

____

____

____

Why was this effect seen? ______________________

____

What was the effect of TSH on the hypophysectomized rat's metabolic rate? How does it compare to the hypophysectomized rat's baseline metabolic rate?

____

____

____

Why was this effect seen? ______________________

____ ■

Activity 4: Determining the Effect of Propylthiouracil on Metabolic Rate

Next, you will investigate the effects of propylthiouracil injections on the metabolic rates of the three rats. Keep in mind that propylthiouracil is a drug that inhibits the production of thyroxine. Select a rat to experiment on first, and then proceed.

1. Under **Data Sets,** highlight **Normal, Tx,** or **Hypox** depending on which rat you are using (**Normal** for the normal rat, **Tx** for thyroidectomized, **Hypox** for hypophysectomized.)

2. Click the **Reset** button in the box labeled *Apparatus*.

3. Click and drag the syringe labeled **propylthiouracil** over to the rat and release the mouse button, injecting the rat.

4. Click and drag the rat into the chamber. Perform steps 1–12 of Activity 1 again, except this time record your data in the With Propylthiouracil section of the chart.

5. Click **Record Data.**

6. Click and drag the rat from the chamber back to its cage and click **Clean** to cleanse the rat of all traces of propylthiouracil.

7. Now repeat this activity for the remaining rats. Record your data in the With Propylthiouracil section of the chart under the corresponding column for each rat.

What was the effect of propylthiouracil on the normal rat's metabolic rate? How does it compare to the normal rat's baseline metabolic rate?

____

____

____

Why was this effect seen? ______________________

____

What was the effect of propylthiouracil on the thyroidectomized rat's metabolic rate? How does it compare to the thyroidectomized rat's baseline metabolic rate?

____

____

____

Why was this effect seen? ______________________

____

What was the effect of propylthiouracil on the hypophysectomized rat's metabolic rate? How does it compare to the hypophysectomized rat's baseline metabolic rate?

____

____

____

Why was this effect seen? ______________________

____

Click **Tools** → **Print Data** to print your recorded data. ■

Hormone Replacement Therapy

Ovaries are stimulated by *follicle-stimulating hormone (FSH)* to get ovarian follicles to develop so that they may be ovulated and fertilized. While the follicles are developing, they produce the hormone *estrogen*. The target tissue for estrogen is the uterus, and the action of estrogen is to enable the uterus to grow and develop so that it may receive fertilized eggs for implantation. *Ovariectomy,* the removal of ovaries, will remove the source of estrogen and cause the uterus to slowly atrophy.

In this activity, you will re-create a classic endocrine experiment and examine how estrogen affects uterine tissue growth. You will be working with two female rats, both of which have been ovariectomized and are thus no longer producing estrogen. You will administer **hormone replacement therapy** to one rat by giving it daily injections of estrogen. The other rat will serve as your control and receive daily injections of saline. You will then remove the uterine tissues from both rats, weigh the tissues, and compare them to determine the effects of hormone replacement therapy.

Start by selecting **Hormone Replacement Therapy** from the **Experiment** menu. A new screen will appear (Figure 4.2), with the two ovariectomized rats in cages. (Note that if this were a wet lab, the ovariectomies would need to have been performed on the rats a month prior to the rest of the experiment in order to ensure that no residual hormones remained in the rats' systems.) Also on screen are a bottle of saline, a bottle of estrogen, a syringe, a box of weighing paper, and a weighing scale.

Proceed carefully with this experiment. Each rat will disappear from the screen once you remove its uterus, and cannot be brought back unless you restart the experiment. This replicates the situation you would encounter if working with live animals: once the uterus is removed, the animal would have to be sacrificed.

Activity 5: Hormone Replacement Therapy

1. Click on the syringe, drag it to the bottle of **saline,** and release the mouse button. The syringe will automatically fill with 1 ml of saline.

2. Drag the syringe to the **control** rat and place the tip of the needle in the rat's lower abdominal area. Injections into this area are considered *interperitoneal* and will quickly be picked up by the abdominal blood vessels. Release the mouse button—the syringe will empty into the rat and automatically return to its holder. Click **Clean** on the syringe holder to clean the syringe of all residue.

3. Click on the syringe again, this time dragging it to the bottle of **estrogen,** and release the mouse button. The syringe will automatically fill with 1 ml of estrogen.

4. Drag the syringe to the **experimental** rat and place the tip of the needle in the rat's lower abdominal area. Release the mouse button—the syringe will empty into the rat and automatically return to its holder. Click **Clean** on the syringe holder to clean the syringe of all residue.

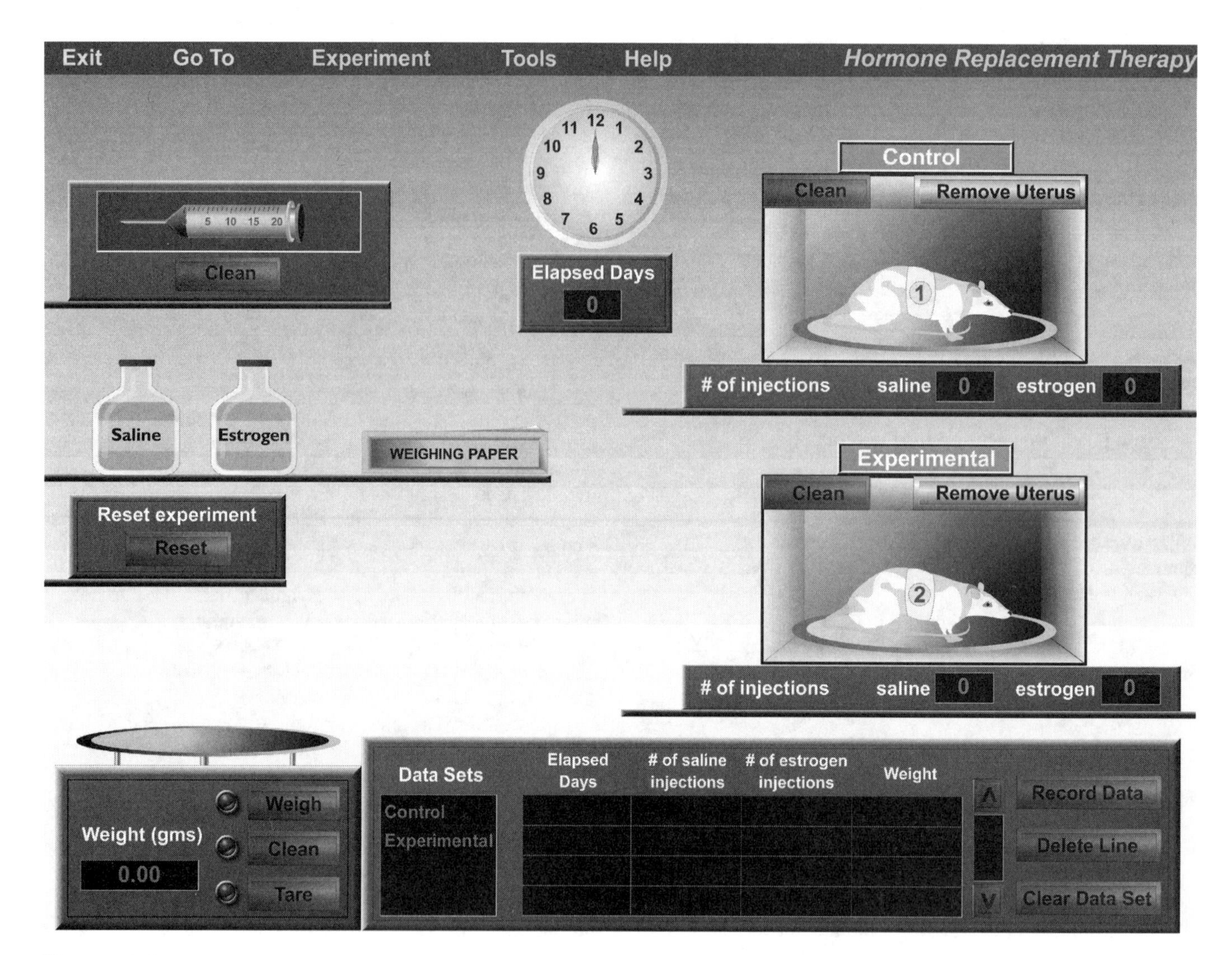

Figure 4.2 Opening screen of the Hormone Replacement Therapy experiment.

5. Click on the **Clock.** You will notice the hands sweep the clock face twice, indicating that 24 hours have passed.

6. Repeat steps 1–5 until each rat has received a total of 7 injections over the course of 7 days (1 injection per day). Note that the **# of injections** displayed below each rat cage records how many injections the rat has received. The control rat should receive 7 injections of saline, whereas the experimental rat should receive 7 injections of estrogen.

7. Next, click on the box of weighing paper. You will see a small piece of paper appear. Click and drag this paper over to the top of the scale and release the mouse button.

8. Notice the scale will give you a weight for the paper. With the mouse arrow, click on the **Tare** button to tare the scale to zero (0.0000 g), adjusting for the weight of the paper.

9. You are now ready to remove the uteruses. In a wet lab, this would require surgery. Here, you will simply click on the **Remove Uterus** button found in each rat cage. The rats will disappear, and a uterus (consisting of a uterine body and two uterine horns) will appear in each cage.

10. Click and drag the uterus from the **control** rat over to the scale and release it on the weighing paper. Click on the **Weigh** button to obtain the weight. Record the weight here:

Uterus weight (control): ______________ g

11. Click **Record Data.**

12. Click **Clean** on the weight scale to dispense of the weighing paper and uterus.

13. Repeat steps 7 and 8. Then click and drag the uterus from the **experimental** rat over to the scale and release it on the weighing paper. Click **Weigh** to obtain the weight. Record the weight here:

Uterus weight (experimental): ______________ g

14. Click **Record Data.**

15. Click **Clean** on the weight scale to dispense of the weighing paper and uterus.

16. Click **Tools → Print Data** to print your recorded data.

How does the control uterus weight compare to the experimental uterus weight?

____

What can you conclude about the administration of estrogen injections on the experimental animal?

____

____

What might be the effect if testosterone had been administered instead of estrogen? Explain your answer.

____

____

____ ■

Insulin and Diabetes

Insulin is produced by the beta cells of the endocrine portion of the pancreas. It is vital to the regulation of blood glucose levels because it enables the body's cells to absorb glucose from the bloodstream. When insulin is not produced by the pancreas, **diabetes mellitus Type I** results. When insulin is produced by the pancreas but the body fails to respond to it, **diabetes mellitus Type II** results. In either case, glucose remains in the bloodstream, unable to be taken up by the body's cells to serve as the primary fuel for metabolism.

In the following experiment, you will study the effects of insulin treatment for type I diabetes. The experiment is divided into two parts. In Part I, you will obtain a *glucose standard curve,* which will be explained shortly. In Part II, you will compare the glucose levels of a normal rat versus that of a diabetic rat, and then compare them again after each rat has been injected with insulin.

Part I

Activity 6:
Obtaining a Glucose Standard Curve

To begin, select **Insulin and Diabetes-Part 1** from the **Experiments** menu (see Figure 4.3).

Select **Balloons On** from the Help menu for help identifying the equipment on-screen. (You will see labels appear as you roll over each piece of equipment.) Select **Balloons Off** to turn this feature off before you begin the experiments.

On the right side of the opening screen is a special spectrophotometer. The **spectrophotometer** is one of the most widely used research instruments in biology. It is used to measure the amounts of light of different wavelengths absorbed and transmitted by a pigmented solution. Inside the spectrophotometer is a source for white light, which is separated into various wavelengths (or colors) by a prism. The user selects a wavelength (color), and light of this color is passed through a tube, or *cuvette,* containing the sample being tested. (For this experiment, the spectrophotometer light source will be preset for a wavelength of 450 nm.) The light transmitted by the sample then passes onto a photoelectric tube, which converts the light energy into an electrical current. The current is then measured by a meter. Alternatively, the light may be measured before the sample is put into the light path, and the amount of light absorbed—called **optical density**—is measured. Using either method, the change in light transmittance or light absorbed can be used to measure the amount of a given substance in the sample being tested.

In Part II, you will use the spectrophotometer to determine how much glucose is present in blood samples taken from two rats. Before using the spectrophotometer, you must obtain a **glucose standard curve** so that you have a point of reference for converting optical density readings into glucose readings, which will be measured in mg/deciliter(mg/dl). To do this, you will prepare five test tubes that contain known amounts of glucose: 30 mg/dl, 60 mg/dl, 90 mg/dl, 120 mg/dl, and 150 mg/dl, respectively. You will then use the spectrophotometer to determine the corresponding optical density readings for each of these known amounts of glucose. Information obtained in Part I will be used to perform Part II.

Also on the screen are three dropper bottles, a test tube washer, a test tube dispenser (on top of the washer), and a test tube incubation unit that you will need to prepare the samples for analysis.

1. Click and drag the test tube (on top of the test tube washer) into slot 1 of the incubation unit. You will see another test tube pop up from the dispenser. Click and drag this second test tube into slot 2 of the incubation unit. Repeat until you have dragged a total of five test tubes into the five slots in the incubation unit.

2. Click and hold the mouse button on the dropper cap of the **glucose standard** bottle. Drag the dropper cap over to tube 1. Release the mouse button to dispense the glucose. You will see that one drop of glucose solution is dropped into the tube and that the dropper cap automatically returns to the bottle of glucose standard.

3. Repeat step 2 with the remaining four tubes. Notice that each subsequent tube will automatically receive one additional drop of glucose standard into the tube (i.e., tube 2 will receive two drops, tube 3 will receive three drops, tube 4 will receive four drops, and tube 5 will receive 5 drops).

4. Click and hold the mouse button on the dropper cap of the **deionized water** bottle. Drag the dropper cap over to tube 1. Release the mouse button to dispense the water. Notice that four drops of water are automatically added to the first tube.

5. Repeat step 4 with tubes 2, 3, and 4. Notice that each subsequent tube will receive one *less* drop of water than the previous tube (i.e., tube 2 will receive three drops, tube 3 will receive two drops, and tube 4 will receive one drop. Tube 5 will receive *no* drops of water.)

6. Click on the **Mix** button of the incubator to mix the contents of the tubes.

7. Click on the **Centrifuge** button. The tubes will descend into the incubator and be centrifuged.

8. When the tubes resurface, click on the **Remove Pellet** button. Any pellets from the centrifuging process will be removed from the test tubes.

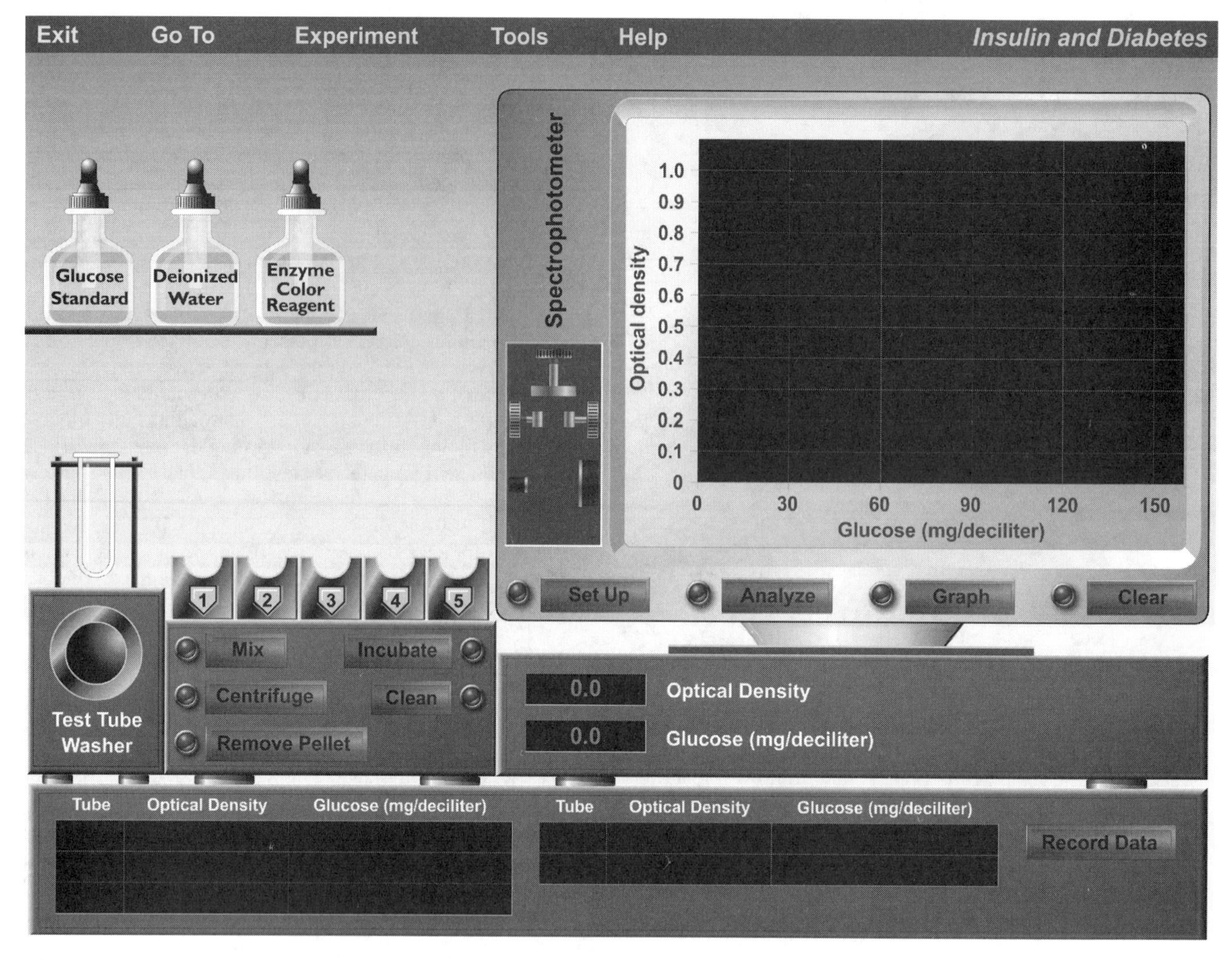

Figure 4.3 Opening screen of the Insulin and Diabetes experiment, Part I.

9. Click and hold the mouse button on the dropper cap of the **enzyme-color reagent** bottle. Still holding the mouse button down, drag the dropper cap over to tube 1. When you release the mouse, you will note that 5 drops of reagent are added to the tube and that the stopper is returned to its bottle.
10. Repeat step 9 for the remaining tubes.
11. Now click **Incubate.** The tubes will descend into the incubator, incubate, and then resurface.
12. Using the mouse, click on **Set Up** on the spectrophotometer. This will warm up the instrument and get it ready for your readings.
13. Click and drag tube 1 into the spectrophotometer (right above the **Set Up** button) and release the mouse button. The tube will lock into place.
14. Click **Analyze.** You will see a spot appear on the screen and values appear in the **Optical Density** and **Glucose** displays.
15. Click **Record Data** on the data collection unit.
16. Click and drag the tube into the test tube washer.
17. Repeat steps 13–16 for the remaining test tubes.
18. When all five tubes have been analyzed, click on the **Graph** button. This is the glucose standard graph, which you will use in Part II of the experiment. Click **Tools → Print Data** to print your recorded data. ■

Part II

Activity 7:

Comparing Glucose Levels Before and After Insulin Injection

Select **Insulin and Diabetes-Part 2** from the **Experiments** menu.

The opening screen will look similar to the screen from Part I (see Figure 4.4). Notice the two rats in their cages. One will be your control animal, the other your experimental animal. Also note the three syringes, containing insulin, saline, and alloxan, respectively. **Alloxan** is a rather nasty drug: when administered to an animal, it will selectively kill all the cells that produce insulin and render the animal instantly diabetic. It does this by destroying the beta cells of the pancreas, which are responsible for insulin production.

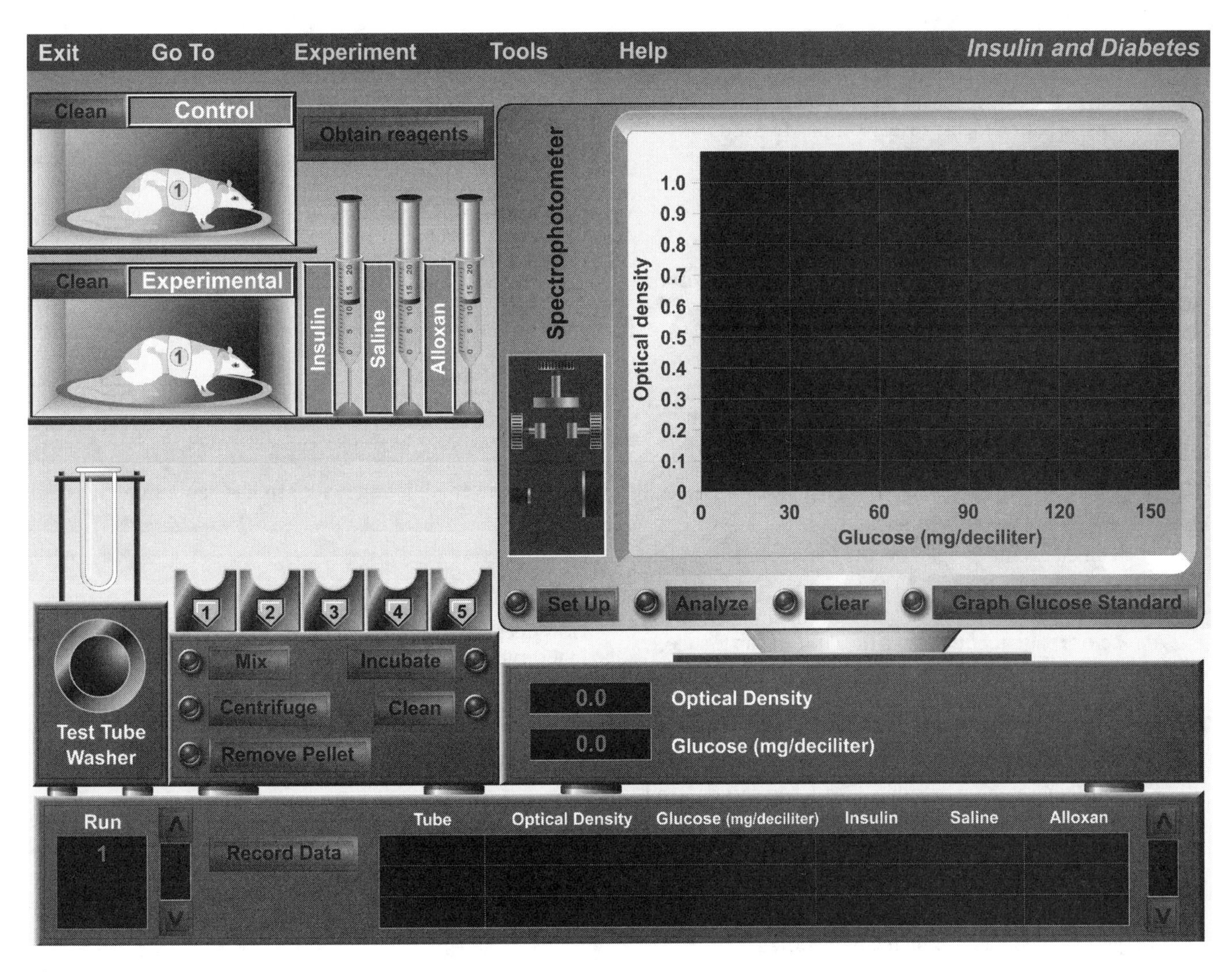

Figure 4.4 Opening screen of the Insulin and Diabetes experiment, Part II.

In this experiment, you will inject the control rat with saline and the experimental rat with alloxan. (Normally, injections are given *daily* for a week. In this simulation, we will administer the injections only once but will be able to see results as though the injections had been given over a longer period of time.)

After administering the saline and alloxan injections, you will obtain blood samples from the two rats. You will then inject both rats with insulin, and obtain blood samples again. Finally, you will analyze all the blood samples in the spectrophotometer (described in Part I) comparing the amounts of glucose present.

1. Click and drag the **saline** syringe to the **control** rat and release the mouse button to inject the animal.

2. Click and drag the **alloxan** syringe to the **experimental** rat and release the mouse button to inject the animal.

3. Click and drag a new test tube (from the test tube dispenser) over to the tail of the **control** rat and release the mouse button. You will note three drops of blood being drawn from the tail into the tube. Next, click and drag the tube into test tube holder **1** in the incubator. (*Note:* The tail is a popular place to get blood from a rat. The end of the tail can easily be clipped and blood collected without really disturbing the rat. The tail heals quickly with no harm to the animal.)

4. Click and drag another new test tube (from the test tube dispenser) over to the tail of the **experimental** rat and release the mouse button. Again, you will note that three drops of blood are drawn from the tail into the tube. Click and drag the tube into test tube holder **2** in the incubator.

5. Click and drag the **insulin** syringe to the **control** rat and release the mouse button to inject the animal.

6. Repeat step 5 with the **experimental** rat.

7. Repeat steps 3 and 4 again, drawing blood samples from each rat and placing the samples into test tube holders **3** and **4.**

8. Click the **Obtain Reagents** button on the cabinet that currently displays the syringes. The syringes will disappear, and you will see four dropper bottles in their place.

9. Click and hold the mouse button on the dropper of the **deionized water** bottle. Drag the dropper cap over to tube 1. Release the mouse button to dispense. You will note that five drops of water are added to the tube.

10. Repeat step 9 for the remaining test tubes.

11. Click and hold the mouse button on the dropper of **barium hydroxide.** Drag the dropper cap over to tube 1. Release the mouse button to dispense. Note that five drops of solution are added to the tube. (Barium hydroxide is used for clearing proteins and cells so that clear glucose readings may be obtained.)

12. Repeat step 11 for the remaining test tubes.

13. Click and hold the mouse button on the dropper of the **heparin** bottle. Still holding the mouse button down, drag the dropper cap over to tube 1. Release the mouse button to dispense.

14. Repeat step 13 for the remaining test tubes.

15. Click on the **Mix** button of the incubator to mix the contents of the tubes.

16. Click on the **Centrifuge** button. The tubes will descend into the incubator, be centrifuged, and then resurface.

17. Click on the **Remove Pellet** button to remove any pellets from the centrifuging process.

18. Click and hold the mouse button on the dropper of the **enzyme-color reagent** bottle. Drag the dropper cap to tube 1. Release the mouse to dispense.

19. Repeat step 18 with the remaining test tubes. In an actual wet lab, you would also shake the test tubes after adding the enzyme-color reagent.

20. Click **Incubate** one more time. The tubes will descend into the incubator, incubate, and then resurface.

21. Click on **Set Up** on the spectrophotometer to warm up the instrument and get it ready for your readings.

22. Click **Graph Glucose Standard.** The graph from Part I of the experiment will appear on the monitor.

23. Click and drag tube 1 to the spectrophotometer and release the mouse button. The tube will lock into place.

24. Click **Analyze.** You will see a horizontal line appear on the screen and a value appear in the **Optical Density** display.

25. Drag the **movable rule** (the vertical line on the far right of the spectrophotometer monitor) over to where the horizontal line (from step 24) crosses the glucose standard line. Watch what happens to the Glucose display as you move the movable rule to the left.

What is the glucose reading where the horizontal line crosses the glucose standard line? Test tube 1: __________ mg/dl glucose

This is your glucose reading for the sample being tested.

26. Click **Record Data** on the data collection unit.

27. Click and drag the test tube from the spectrophotometer into the test tube washer, then click **Clear** underneath the oscilloscope display.

28. Repeat steps 22–27 for the remaining test tubes. Record your glucose readings for each test tube here:

Test tube 2: __________ mg/dl glucose

Test tube 3: __________ mg/dl glucose

Test tube 4: __________ mg/dl glucose

How does the glucose level in test tube 1 compare to the level in test tube 2? Recall that tube 1 contains a sample from your control rat (which received injections of saline) and that tube 2 contains a sample from your experimental rat (which received injections of alloxan).

____

Explain this result: ______________________________

____

What is the condition that alloxan has caused in the experimental rat?

____

How does the glucose level in test tube 3 compare to the level in test tube 1?

____

Explain this result: ______________________________

____

How does the glucose level in test tube 4 compare to the level in test tube 2?

____

Explain this result: ______________________________

____

What was the effect of administering insulin to the control animal?

____

What was the effect of administering insulin to the experimental animal?

____

Click **Tools** → **Print Data** to print your recorded data. ■

Histology Review Supplement

Turn to p. P-144 for a review of endocrine tissue.

Cardiovascular Dynamics: Computer Simulation

Exercise 5

Objectives

1. To define the following: *blood flow; peripheral resistance; viscosity; systole; diastole; end diastolic volume; end systolic volume; stroke volume; cardiac output.*
2. To explore cardiovascular dynamics using an experimental setup to simulate a human body function.
3. To understand that heart and blood vessel functions are highly coordinated.
4. To comprehend that pressure differences provide the driving force that moves blood through the blood vessels.
5. To recognize that body tissues may differ in their blood demands at a given time.
6. To identify the most important factors in control of blood flow.
7. To comprehend that changing blood vessel diameter can alter the pumping ability of the heart.
8. To examine the effect of stroke volume on blood flow.

The physiology of human blood circulation can be divided into two distinct but remarkably harmonized processes: (1) the pumping of blood by the heart, and (2) the transport of blood to all body tissues via the vasculature, or blood vessels. Blood supplies all body tissues with the substances needed for survival, so it is vital that blood delivery is ample for tissue demands.

The Mechanics of Circulation

To understand how blood is transported throughout the body, let's examine three important factors influencing how blood circulates through the cardiovascular system: blood flow, blood pressure, and peripheral resistance.

Blood flow is the amount of blood moving through a body area or the entire cardiovascular system in a given amount of time. While total blood flow is determined by cardiac output (the amount of blood the heart is able to pump per minute), blood flow to specific body areas can vary dramatically in a given time period. Organs differ in their requirements from moment to moment, and blood vessels constrict or dilate to regulate local blood flow to various areas in response to the tissue's immediate needs. Consequently, blood flow can increase to some regions and decrease to other areas at the same time.

Blood pressure is the force blood exerts against the wall of a blood vessel. Owing to cardiac activity, pressure is highest at the heart end of any artery. Because of the effect of peripheral resistance, which will be discussed shortly, pressure within the arteries (or any blood vessel) drops as the distance (vessel length) from the heart increases. This pressure gradient causes blood to move from and then back to the heart, always moving from high- to low-pressure areas.

Peripheral resistance is the opposition to blood flow resulting from the friction developed as blood streams through blood vessels. Three factors affect vessel resistance: blood viscosity, vessel radius, and vessel length.

Blood viscosity is a measure of the "thickness" of the blood, and is caused by the presence of proteins and formed elements in the plasma (the fluid part of the blood). As the viscosity of a fluid increases, its flow rate through a tube decreases. Blood viscosity in healthy people normally does not change, but certain conditions such as too many or too few blood cells may modify it.

Controlling *blood vessel radius* (one-half of the diameter) is the principal method of blood flow control. This is accomplished by contracting or relaxing the smooth muscle within the blood vessel walls. To see why radius has such a pronounced effect on blood flow, consider the physical relationship between blood and the vessel wall. Blood in direct contact with the vessel wall flows relatively slowly because of the friction, or drag, between the blood and the lining of the vessel. In contrast, fluid in the center of the vessel flows more freely because it is not rubbing against the vessel wall. Now picture a large- and a small-radius vessel: proportionately more blood is in contact with the wall of the small vessel; hence blood flow is notably impeded as vessel radius decreases.

Although *vessel length* does not ordinarily change in a healthy person, any increase in vessel length causes a corresponding flow decrease. This effect is principally caused by friction between blood and the vessel wall. Consequently, given two blood vessels of the same diameter, the longer vessel will have more resistance, and thus a reduced blood flow.

The Effect of Blood Pressure and Vessel Resistance on Blood Flow

Poiseuille's equation describes the relationship between pressure, vessel radius, viscosity, and vessel length on blood flow:

$$\text{Blood flow } (\Delta Q) = \frac{\pi \Delta P\, r^4}{8 \eta l}$$

In the equation, ΔP is the pressure difference between the two ends of the vessel and represents the driving force behind blood flow. Viscosity (η) and blood vessel length (l) are not commonly altered in a healthy adult. We can also see from the equation that blood flow is directly proportional to the fourth power of vessel radius (r^4), which means that small variations in vessel radius translate into large changes in blood flow. In the human body, changing blood vessel radius provides an extremely effective and sensitive method of blood flow control. Peripheral resistance is the most important factor in blood flow control, because circulation to individual organs can be independently regulated even though systemic pressure may be changing.

Vessel Resistance

Imagine for a moment that you are one of the first cardiovascular researchers interested in the physics of blood flow. Your first task as the principal investigator for this project is to plan an effective experimental design simulating a simple fluid pumping system that can be related to the mechanics of the cardiovascular system. The initial phenomenon you study is how fluids, including blood, flow through tubes or blood vessels. Questions you might ask include:

1. What role does pressure play in the flow of fluid?
2. How does peripheral resistance affect fluid flow?

The equipment required to solve these and other questions has already been designed for you in the form of a computerized simulation, which frees you to focus on the logic of the experiment. The first part of the computer simulation indirectly investigates the effects of pressure, vessel radius, viscosity, and vessel length on fluid flow. The second part of the experiment will explore the effects of several variables on the output of a single-chamber pump. Follow the specific guidelines in the exercise for collecting data. As you do so, also try to imagine alternate methods of achieving the same experimental goal.

Choose **Cardiovascular Dynamics** from the main menu. The opening screen will appear in a few seconds (Figure 5.1).

The primary features on the screen when the program starts are a pair of glass beakers perched atop a simulated electronic device called the *equipment control unit,* which is used to set experiment parameters and to operate the equipment. When the **Start** button (beneath the left beaker) is clicked, the simulated blood flows from the left beaker (source) to the right beaker (destination) through the connecting tube. To relate this to the human body, think of the left beaker as the left side of your heart, the tube as your aorta, and the right beaker as any organ to which blood is flowing.

Clicking the **Refill** button refills the source beaker after an experimental trial. Experimental parameters can be adjusted by clicking the plus (+) or minus (−) buttons to the right of each display window.

The equipment in the lower part of the screen is called the *data collection unit*. This equipment records and displays data you accumulate during the experiments. The data set for the first experiment (Radius) is highlighted in the **Data Sets** window. You can add or delete a data set by clicking the appropriate button to the right of the **Data Sets** window. The **Record Data** button in the lower right part of the screen activates automatically after an experimental trial. Clicking the **Delete Line** or **Clear Data Set** buttons erases any data you want to delete.

You will record the data you accumulate in the experimental values grid in the lower middle part of the screen.

Activity 1: Studying the Effect of Flow Tube Radius on Fluid Flow

Our initial study will examine the effect of flow tube radius on fluid flow.*

1. Open the Vessel Resistance window if it isn't already open.

The Radius line in the data collection unit should be highlighted in bright blue. If it is not, choose it by clicking the **Radius** line. The data collection unit will now record flow variations due to changing flow tube radius.

* If you need help identifying any piece of equipment, choose Balloons On from the Help menu and move the mouse pointer onto any piece of equipment visible on the computer's screen. As the pointer touches the object, a pop-up window appears to identify the equipment. To close the pop-up window, move the mouse pointer away from the equipment. Repeat the process for all equipment on the screen until you feel confident with it. When finished, choose Balloons off to turn off this help feature.

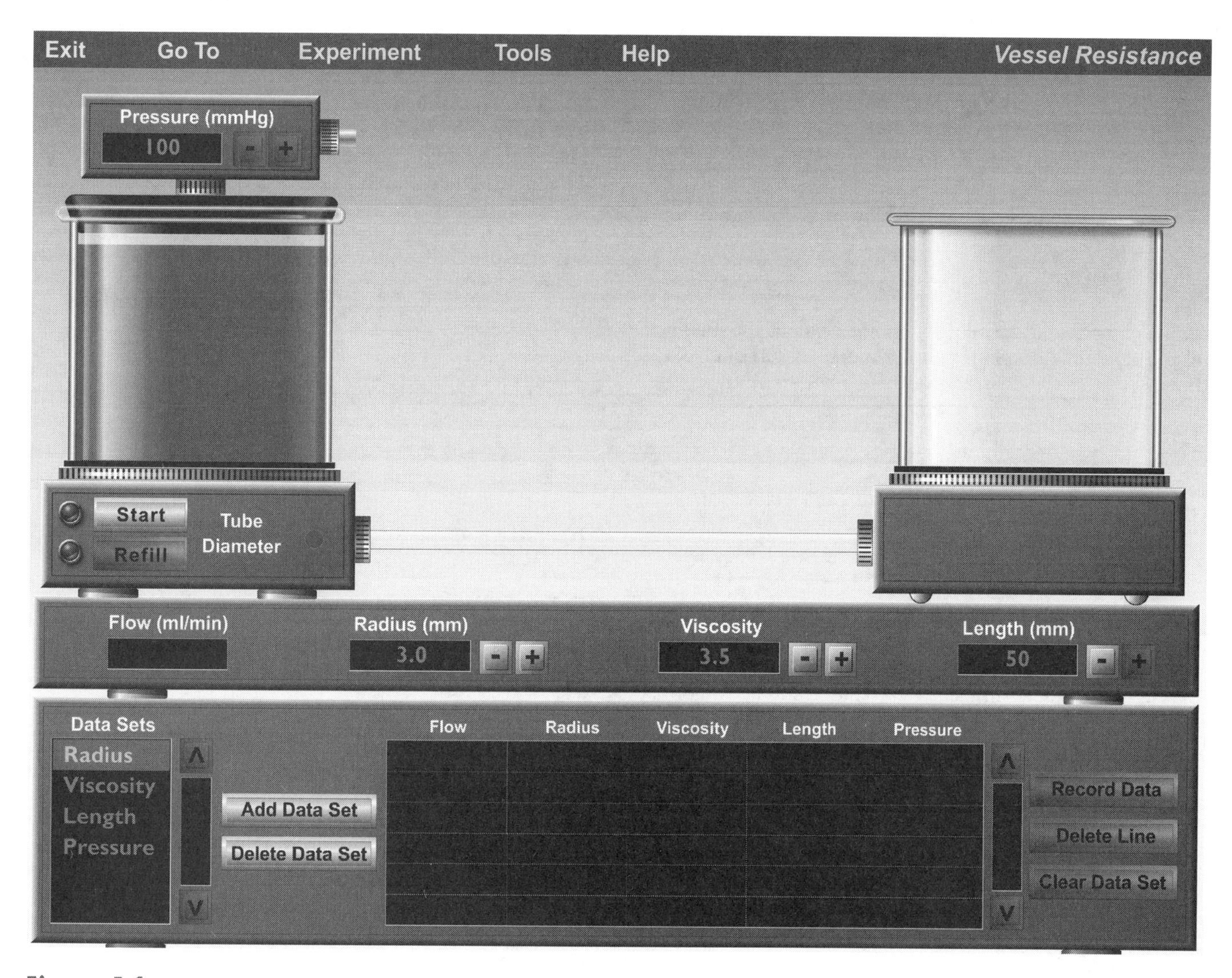

Figure 5.1 Opening screen of the Vessel Resistance experiment.

If the data grid is not empty, click **Clear Data Set** to discard all previous values.

If the left beaker is not full, click **Refill.**

2. Adjust the flow tube radius to 1.5 mm and the viscosity to 1.0 by clicking the appropriate (+) or (−) button. During the course of this part of the experiment, maintain the other experiment conditions at:

100 mm Hg driving pressure (top left)

50-mm flow tube length (middle right)

3. Click **Start** and watch the fluid move into the right beaker. (Fluid moves slowly under some conditions—be patient!) Pressure (currently set to 100 mm Hg) propels fluid from the left beaker to the right beaker through the flow tube. The flow rate is displayed in the Flow window when the left beaker has finished draining. Now click **Record Data** to display the flow rate and experiment parameters in the experimental values grid (and retain the data in the computer's memory for printing and saving). Click **Refill** to replenish the left beaker.

4. Increase the radius in 0.5-mm increments and repeat step 3 until the maximum radius (6.0 mm) is achieved. Be sure to click **Record Data** after each fluid transfer. If you make an error and want to delete a data value, click the data line in the experimental values grid and then click **Delete Line.**

5. View your data graphically by choosing **Plot Data** from the **Tools** menu. Choose **Radius** as the data set to be graphed, and then use the slider bars to select the radius data to be plotted on the X-axis and the flow data to be plotted on the Y-axis. You can highlight individual data points by clicking a line in the data grid. When you are finished, click the close box at the top of the plot window.

What happened to fluid flow as the radius of the flow tube was increased?

____

____

____

Because fluid flow is proportional to the fourth power of the radius, ____________ changes in tube radius cause ____________ changes in fluid flow.

Is the relationship between fluid flow and flow tube radius linear or exponential? ____________

In this experiment, a simulated motor changes the diameter of the flow tube. Explain how our blood vessels alter blood flow.

____

____

____

____

After a heavy meal when we are relatively inactive, we might expect blood vessels in the skeletal muscles to be somewhat ____________, whereas blood vessels in the digestive organs are probably ____________. ■

Activity 2: Studying the Effect of Viscosity on Fluid Flow

With a viscosity of 3 to 4, blood is much more viscous than water (1.0 viscosity). Although viscosity is altered by factors such as dehydration and altered blood cell numbers, a body in homeostatic balance has a relatively stable blood consistency. Nonetheless it is useful to examine the effects of viscosity changes on fluid flow, because we can then predict what might transpire in the human cardiovascular system under certain homeostatic imbalances.

1. Set the starting conditions as follows:

- 100 mm Hg driving pressure
- 5.0-mm flow tube radius
- 1.0 viscosity
- 50-mm flow tube length

2. Click the **Viscosity** data set in the data collection unit. (This action prepares the experimental values grid to record the viscosity data.)

3. **Refill** the left beaker if you have not already done so.

4. Click **Start** to begin the experiment. After all the fluid has drained into the right beaker, click **Record Data** to record this data point, and then click **Refill** to replenish the left beaker.

5. In 1.0-unit increments, increase the fluid viscosity and repeat step 4 until the maximum viscosity (10.0) is reached.

6. View your data graphically by choosing **Plot Data** from the **Tools** menu. Choose **Viscosity** as the data set to be graphed, and then use the slider bars to select the viscosity data to be plotted on the X-axis and the flow data to be plotted on the Y-axis. You can highlight individual data points by clicking a line in the data grid. When finished, click the close box at the top of the plot window.

How does fluid flow change as viscosity is modified?

____

____

____

Is fluid flow versus viscosity an inverse or direct relationship?

____

How does the effect of viscosity compare with the effect of radius on fluid flow?

____

____

Predict the effect of anemia (e.g., fewer red blood cells than normal) on blood flow.

____

____

What might happen to blood flow if we increased the number of blood cells?

____

____

____

Explain why changing blood viscosity would or would not be a reasonable method for the body to control blood flow.

____

____

____

____

Blood viscosity would ____________ in conditions of dehydration, resulting in ____________ blood flow. ■

Activity 3:
Studying the Effect of Flow Tube Length on Fluid Flow

With the exception of the normal growth that occurs until the body reaches full maturity, blood vessel length does not significantly change. In this activity you will investigate the physical relationship between vessel length and blood movement; specifically, how blood flow changes in flow tubes (vessels) of constant radius but of different lengths.

1. Set the starting conditions as follows:

- 100 mm Hg driving pressure
- 5.0-mm flow tube radius
- 3.5 viscosity
- 10-mm flow tube length

2. Click the **Length** data set in the data collection unit. (This action prepares the experimental values grid to display the length data.)

3. **Refill** the left beaker if you have not already done so.

4. Click **Start** to begin the experiment. After all the fluid has drained into the right beaker, click **Record Data** to record this data point, and then click **Refill** to refill the left beaker.

5. In 5-mm increments, increase the flow tube length by clicking (+) next to the Length window and repeat step 4 until the maximum length (50 mm) has been reached.

6. View your data graphically by choosing **Plot Data** from the **Tools** menu. Choose **Length** as the data set to be graphed, and then use the slider bars to select the length data to be plotted on the X-axis and the flow data to be plotted on the Y-axis. You can highlight individual data points by clicking a line in the grid. When finished, click the close box at the top of the plot window.

How does flow tube length affect fluid flow?

____

____

____

Explain why altering blood vessel length would or would not be a good method of controlling blood flow in the body.

____

____

____

______________________________________ ■

Activity 4:
Studying the Effect of Pressure on Fluid Flow

The pressure difference between the two ends of a blood vessel is the driving force behind blood flow. In comparison, our experimental setup pressurizes the left beaker, thereby providing the driving force that propels fluid through the flow tube to the right beaker. You will examine the effect of pressure on fluid flow in this part of the experiment.

1. Set the starting conditions as follows:

- 25 mm Hg driving pressure
- 5.0-mm flow tube radius
- 3.5 viscosity
- 50-mm flow tube length

2. Click the **Pressure** data set in the data collection unit. (This action prepares the experimental values grid for the pressure data.)

3. **Refill** the left beaker if you have not already done so.

4. Click **Start** to begin the experiment. After all the fluid has moved into the right beaker, click **Record Data** to record this data point. Click **Refill** to refill the left beaker.

5. In increments of 25 mm Hg, increase the driving pressure by clicking (+) next to the Pressure window and repeat step 4 until the maximum pressure (225 mm Hg) has been reached.

6. View your data graphically by choosing **Plot Data** from the **Tools** menu. Choose **Pressure** as the data set to be graphed, and then use the slider bars to select the pressure data to be plotted on the X-axis and the flow data to be plotted on the Y-axis. You can highlight individual data points by clicking a line in the data grid. When finished, click the close box at the top of the plot window.

How does driving pressure affect fluid flow?

____

____

____

How does this plot differ from the others?

____

____

____

Although changing pressure could be used as a means of blood flow control, explain why this approach would not be as effective as altering blood vessel radius.

____

____

____

____

Click **Tools → Print Data** to print your recorded data. ■

Pump Mechanics

In the human body, the heart beats approximately 70 strokes each minute. Each heartbeat consists of a filling interval during which blood moves into the chambers of the heart, and an ejection period when blood is actively pumped into the great arteries. The pumping activity of the heart can be described in terms of the phases of the cardiac cycle. Heart chambers fill during **diastole** (relaxation of the heart) and pump blood out during **systole** (contraction of the heart). As you can imagine, the length of time the heart is relaxed is one factor that determines the amount of blood within the heart at the end of the filling interval. Up to a point, increasing ventricular filling time results in a corresponding increase in ventricular volume. The volume in the ventricles at the end of diastole, just before cardiac contraction, is called the **end diastolic volume,** or **EDV.**

Blood moves from the heart into the arterial system when systolic pressure increases above the residual pressure (from the previous systole) in the great arteries leaving the heart. Although ventricular contraction causes blood ejection, the heart does not empty completely; a small quantity of blood—the **end systolic volume,** or **ESV**—remains in the ventricles at the end of systole.

Because the oxygen requirements of body tissue change depending on activity levels, we would expect the **cardiac output** (amount of blood pumped by each ventricle per minute) to vary correspondingly. We can calculate the **stroke volume** (**SV,** the amount of blood pumped per contraction of each ventricle) by subtracting the end systolic volume from the end diastolic volume (SV = EDV – ESV). We then compute cardiac output by multiplying the stroke volume by heart rate.

The human heart is a complex four-chambered organ, consisting of two individual pumps (the right and left sides) connected together in series. The right heart pumps blood through the lungs into the left heart, which in turn delivers blood to the systems of the body. Blood then returns to the right heart to complete the circuit.

Using the Pump Mechanics part of the program you will explore the operation of a simple one-chambered pump, and apply the physical concepts in the simulation to the operation of either of the two pumps composing the human heart.

In this experiment you can vary the starting and ending volumes of the pump (analogous to EDV and ESV, respectively), driving and resistance pressures, and the diameters of the flow tubes leading to and from the pump chamber. As you proceed through the exercise, try to apply the ideas of ESV, EDV, cardiac output, stroke volume, and blood flow to the on-screen simulated pump system. For example, imagine that the flow tube leading to the pump from the left represents the pulmonary veins, while the flow tube exiting the pump to the right represents the aorta. The pump would then represent the left side of the heart.

Select **Pump Mechanics** from the **Experiment** menu. The equipment for the Pump Mechanics part of the experiment appears (Figure 5.2).

As in the previous experiment, there are two simulated electronic control units on the computer's screen. The upper apparatus is the *equipment control unit,* which is used to adjust experiment parameters and to operate the equipment. The lower apparatus is the *data collection* unit, in which you will record the data you accumulate.

This equipment differs slightly from that used in the Vessel Resistance experiment. There are still two beakers: the *source* beaker on the left and the *destination* beaker on the right. Now the pressure in each beaker is individually controlled by the small pressure units on top of the beakers. Between the two beakers is a simple pump, which can be thought of as one side of the heart, or even as a single ventricle (e.g., left ventricle). The left beaker and flow tube are analogous to the venous side of human blood flow, while arterial circulation is simulated by the right flow tube and beaker. One-way valves in the flow tubes supplying the pump ensure fluid movement in one direction—from the left beaker into the pump, and then into the right beaker. If you imagine that the pump represents the left ventricle, then think of the valve to the left of the pump as the bicuspid valve and the valve to the right of the pump as the aortic semilunar valve. The pump is driven by a pressure unit mounted on its cap. An important distinction between the pump's pressure unit and the pressure units atop the beakers is that the pump delivers pressure only during its downward stroke. Upward pump strokes are driven by pressure from the left beaker (the pump does not exert any resistance to flow from the left beaker during pump filling). In contrast, pressure in the right beaker works against the pump pressure, which means that the net pressure driving the fluid into the right beaker is calculated (automatically) by subtracting the right beaker pressure from the pump pressure. The resulting pressure difference between the pump and the right beaker is displayed in the experimental values grid in the data collection unit as **Pres.Dif.R.**

Clicking the **Auto Pump** button will cycle the pump through the number of strokes indicated in the Max. strokes window. Clicking the **Single** button cycles the pump through one stroke. During the experiment, the pump and flow rates are automatically displayed when the number of pump strokes is 5 or greater. The radius of each flow tube is individually controlled by clicking the appropriate button. Click (+) to increase flow tube radius or (−) to decrease flow tube radius.

The pump's stroke volume (the amount of fluid ejected in one stroke) is automatically computed by subtracting its ending volume from the starting volume. You can adjust starting and ending volumes, and thereby stroke volume, by clicking the appropriate (+) or (−) button in the equipment control unit.

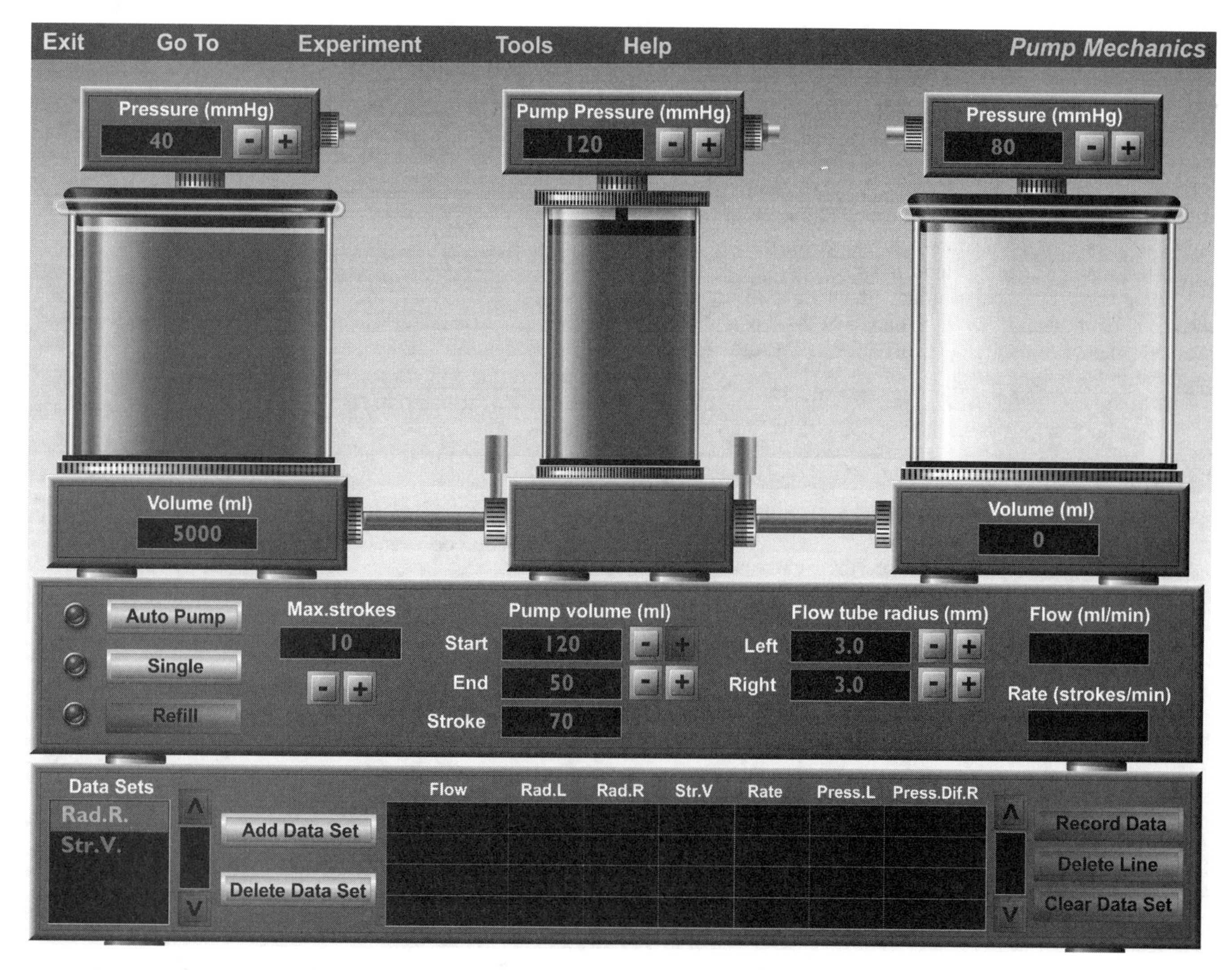

Figure 5.2 Opening screen of the Pump Mechanics experiment.

The data collection unit records and displays data you accumulate during the experiments. The data set for the first experiment (**Rad.R.,** which represents right flow tube radius) is highlighted in the **Data Sets** window. You can add or delete a data set by clicking the appropriate button to the right of the **Data Sets** window. Clicking **Delete Data Set** will erase the data set itself, including all the data it contained. The **Record Data** button at the right edge of the screen activates automatically after an experimental trial. When clicked, the **Record Data** button displays the flow rate data in the experimental values grid and saves it in the computer's memory. Clicking the **Delete Line** or **Clear Data Set** button erases any data you want to delete.

Activity 5: Studying the Effect of Radius on Pump Activity

Although you will be manipulating the radius of only the right flow tube in this part of the exercise, try to predict the consequence of altering the left flow tube radius as you collect the experimental data. (Remember that the left flow tube simulates the pulmonary veins and the right flow tube simulates the aorta.)

1. Open the pump mechanics window if it isn't already open.

Click the **Rad.R.** data set to activate it. The data collection unit is now ready to record flow variations due to changing flow tube radius.

If the experimental values grid is not empty, click **Clear Data Set** to discard all previous values.

If the left beaker is not full, click **Refill.**

2. Adjust the **right** flow tube radius to 2.5 mm, and the **left** flow tube radius to 3.0 mm by clicking and holding the appropriate button. During the entire radius part of the experiment, maintain the other experiment conditions at:

- 40 mm Hg for left beaker pressure (this pressure drives fluid into the pump, which offers no resistance to filling)
- 120 mm Hg for pump pressure (pump pressure is the driving force that propels fluid into the right beaker)

- 80 mm Hg for right beaker pressure (this pressure is the resistance to the pump's pressure)
- 120-ml Start volume in pump (analogous to EDV)
- 50-ml End volume in pump (analogous to ESV)
- 10 strokes in the Max. strokes window

Notice that the displayed 70-ml stroke volume is automatically calculated. Before starting, click the **Single** button one or two times and watch the pump action.

To be sure you understand how this simple mechanical pump can be thought of as a simulation of the human heart, complete the following statements by circling the correct term within the parentheses:

a. When the piston is at the bottom of its travel, the volume remaining in the pump is analogous to (EDV, ESV) of the heart.

b. The amount of fluid ejected into the right beaker by a single pump cycle is analogous to (stroke volume, cardiac output) of the heart.

c. The volume of blood in the heart just before systole is called (EDV, ESV), and is analogous to the volume of fluid present in the simulated pump when it is at the (top, bottom) of its stroke.

3. Click the **Auto Pump** button in the equipment control unit to start the pump. After the 10 stroke volumes have been delivered, the Flow and Rate windows will automatically display the experiment results. Now click **Record Data** to display the figures you just collected in the experimental values grid. Click **Refill** to replenish the left beaker.

4. Increase the *right* flow tube radius in 0.5-mm increments and repeat step 3 above until the maximum radius (6.0 mm) is achieved. Be sure to click **Record Data** after each trial.

5. When you have completed the experiment, view your data graphically by choosing **Plot Data** from the **Tools** menu. Choose **Rad.R.** as the data set to be graphed, and then use the slider bars to select the Rad.R. data to be plotted on the X-axis and the flow data to be plotted on the Y-axis. You can highlight individual data points by clicking a line in the data grid.

The total flow rates you just determined depend on the flow rate into the pump from the left and on the flow rate out of the pump toward the right. Consequently, the shape of the plot is different from what you might predict after viewing the vessel resistance radius graph.

Try to explain why this graph differs from the radius plot in the Vessel Resistance experiment. Remember that the flow rate into the pump did not change, whereas the flow rate out of the pump varied according to your radius manipulations. When you have finished, click the close box at the top of the plot window.

____

____

____

____

Complete the following statements by circling the correct term within the parentheses.

a. As the right flow tube radius is increased, fluid flow rate (increases, decreases). This is analogous to (dilation, constriction) of blood vessels in the human body.

b. Even though the pump pressure remains constant, the pump rate (increases, decreases) as the radius of the right flow tube is increased. This happens because the resistance to fluid flow is (increased, decreased).

Apply your observations of the simulated mechanical pump to complete the following statements about human heart function. If you are not sure how to formulate your response, use the simulation to arrive at an answer. Circle the correct term within the parentheses.

c. The heart must contract (more, less) forcefully to maintain cardiac output if the resistance to blood flow in the vessels exiting the heart is increased.

d. Increasing the resistance (that is constricting) the blood vessels entering the heart would (increase, decrease) the time needed to fill the heart chambers.

What do you think would happen to the flow rate and the pump rate if the left flow tube radius is changed (either increased or decreased)?

____

____

____ ■

Activity 6:
Studying the Effect of Stroke Volume on Pump Activity

Whereas the heart of a person at rest pumps about 60% of the blood in its chambers, 40% of the total amount of blood remains in the chambers after systole. The 60% of blood ejected by the heart is called the stroke volume, and is the difference between EDV and ESV. Even though our simple pump in this experiment does not work exactly like the human heart, you can apply the concepts to basic cardiac function. In this experiment, you will examine how the activity of the simple pump is affected by changing the pump's starting and ending volumes.

1. Click the **Str.V.** (stroke volume) data set to activate it. If the experimental values grid is not empty, click **Clear Data Set** to discard all previous values. If the left beaker is not full, click **Refill.**

2. Adjust the stroke volume to 10 ml by setting the Start volume (EDV) to 120 ml and the End volume (ESV) to 110 ml (stroke volume = start volume – end volume). During the entire stroke volume part of the experiment, keep the other experimental conditions at:

- 40 mm Hg for left beaker pressure
- 120 mm Hg for pump pressure
- 80 mm Hg for right beaker pressure
- 3.0 mm for left and right flow tube radius
- 10 strokes in the Max. strokes window

3. Click the **Auto Pump** button to start the experiment. After 10 stroke volumes have been delivered, the Flow and Rate windows will display the experiment results given the current parameters. Click **Record Data** to display the figures you just collected in the experimental values grid. Click **Refill** to replenish the left beaker.

4. Increase the stroke volume in 10-ml increments (by decreasing the End volume) and repeat step 3 until the maximum stroke volume (120 ml) is achieved. Be sure to click the **Record Data** button after each trial. Watch the pump action during each stroke to see how you can apply the concepts of starting and ending pump volumes to EDV and ESV of the heart.

5. View your data graphically by choosing **Plot Data** from the **Tools** menu. Choose **Str.V.** as the data set to be graphed, and then use the slider bars to select the Str.V. data to be plotted on the X-axis and the pump rate data to be plotted on the Y-axis. Answer the following questions. When you have finished, click the close box at the top of the plot window.

What happened to the pump's rate as its stroke volume was increased?

____

____

____

Using your simulation results as a basis for your answer, explain why an athlete's resting heart rate might be lower than that of the average person.

____

____

____

Applying the simulation outcomes to the human heart, predict the effect of increasing the stroke volume on cardiac output (at any given rate).

____

____

____

When heart rate is increased, the time of ventricular filling is (circle one: increased, decreased), which in turn (increases, decreases) the stroke volume.

What do you think might happen to the pressure in the pump during filling if the valve in the right flow tube became leaky? (Remember that the pump offers no resistance to filling.)

____

____

Applying this concept to the human heart, what might occur in the left heart and pulmonary blood vessels if the aortic valve became leaky?

____

____

What might occur if the aortic valve became slightly constricted?

____

____

In your simulation, increasing the pressure in the right beaker is analogous to the aortic valve becoming (circle one: leaky, constricted). ■

Activity 7:
Studying Combined Effects

In this section, you will set up your own experimental conditions to answer the following questions. Carefully examine each question and decide how to set experiment parameters to arrive at an answer. You can examine your previously collected data if you need additional information. Record several data points for each question as evidence for your answer (unless the question calls for a single pump stroke).

Click the **Add Data Set** button in the data collection unit. Next, create a new data set called Combined. Your newly created data set will be displayed beneath Str.V. Now click the **Combined** line to activate the data set. As you collect the supporting data for the following questions, be sure to click **Record Data** each time you have a data point you wish to keep for your records.

How is the flow rate affected when the right flow tube radius is kept constant (at 3.0 mm) and the left flow tube radius is modified (either up or down)?

____

____

____

____

How does decreasing left flow tube radius affect pump chamber filling time? Does it affect pump chamber emptying?

____

____

____

____

You have already examined the effect of changing the pump's end volume as a way of manipulating stroke volume. What happens to flow and pump rate when you keep the end volume constant and alter the start volume to manipulate stroke volume?

____

____

____

____

Try manipulating the pressure delivered to the left beaker. How does changing the left beaker pressure affect flow rate? (This change would be similar to changing pulmonary vein pressure.)

____

____

If the left beaker pressure is decreased to 10 mm Hg, how is pump-filling time affected?

____

____

____

What happens to the pump rate if the filling time is shortened?

____

____

What happens to fluid flow when the right beaker pressure equals the pump pressure?

____

____

____ ■

Activity 8:
Studying Compensation

In this activity you will explore the concept of cardiovascular compensation. Click the **Add Data Set** button in the data collection unit. Next, create a new data set called **Comp.** Your newly created data set will be displayed beneath Combined. Now click the **Comp. Line** to activate the data set. As you collect the supporting data for the following questions, be sure to click **Record Data** each time you have a data point you wish to keep for your records.

Adjust the experimental conditions to the following:

- 40 mm Hg for left beaker pressure
- 120 mm Hg for pump pressure
- 80 mm Hg for right beaker pressure
- 3.0 mm for left and right flow tube radius
- 10 strokes in the Max. strokes window
- 120-ml Start volume in pump
- 50-ml End volume in pump

Click **Auto Pump** and then record your flow rate data. Let's declare the value you just obtained to be the "normal" flow rate for the purpose of this exercise.

Now decrease the right flow tube radius to 2.5 mm, and run another trial. How does this flow rate compare with "normal"?

____

____

Leave the right flow tube radius at 2.5 mm radius and try to adjust one or more other conditions to return flow to "normal." Think logically about what condition(s) might compensate for a decrease in flow tube radius. How were you able to accomplish this? (Hint: There are several ways.)

____

____

____

Decreasing the right flow tube radius is similar to a partial (circle one: leakage, blockage) of the aortic valve or (increased, decreased) resistance in the arterial system.

Explain how the human heart might compensate for such a condition.

____

____

____

____

To increase (or decrease) blood flow to a particular body system (e.g., digestive), would it be better to adjust heart rate or blood vessel diameter? Explain.

____

____

____

____

Complete the following statements by circling the correct response. (If necessary, use the pump simulation to help you with your answers.)

a. If we decreased overall peripheral resistance in the human body (as in an athlete), the heart would need to generate (more, less) pressure to deliver an adequate amount of blood flow, and arterial pressure would be (higher, lower).

b. If the diameter of the arteries of the body were partly filled with fatty deposits, the heart would need to generate (more, less) force to maintain blood flow, and pressure in the arterial system would be (higher, lower) than normal. ■

Click **Tools** → **Print Data** to print your recorded data.

Histology Review Supplement

Turn to p. P-146 for a review of cardiovascular tissue.

Exercise

6 Frog Cardiovascular Physiology: Computer Simulation

Objectives

1. To list the properties of cardiac muscle as automaticity and rhythmicity and define each.
2. To explain the statement "Cardiac muscle has an intrinsic ability to beat."
3. To compare the relative length of the refractory period of cardiac muscle with that of skeletal muscle, and explain why it is not possible to tetanize cardiac muscle.
4. To define *extrasystole* and to explain at what point in the cardiac cycle (and on an ECG tracing) an extrasystole can be induced.
5. To describe the effect of the following on heart rate: cold, heat, vagal stimulation, pilocarpine, digitalis, atropine, epinephrine, and potassium, sodium, and calcium ions.
6. To define *vagal escape* and discuss its value.
7. To define *ectopic pacemaker.*

Investigation of human cardiovascular physiology is very interesting, but many areas obviously do not lend themselves to experimentation. It would be tantamount to murder to inject a human subject with various drugs to observe their effects on heart activity, or to expose the human heart in order to study the length of its refractory period. However, this type of investigation can be done on frogs or computer simulations, and provides valuable data because the physiological mechanisms in these animals, or programmed into the computer simulation, are similar if not identical to those in humans.

In this exercise, you will conduct the cardiac investigations just mentioned.

Special Electrical Properties of Cardiac Muscle: Automaticity and Rhythmicity

Cardiac muscle differs from skeletal muscle both functionally and in its fine structure. Skeletal muscle must be electrically stimulated to contract. In contrast, heart muscle can and does depolarize spontaneously in the absence of external stimulation. This property, called **automaticity,** is due to plasma membranes that have reduced permeability to potassium ions, but still allow sodium ions to slowly leak into the cells. This leakage causes the muscle cells to slowly depolarize until the action potential threshold is reached and *fast calcium channels* open, allowing Ca^{2+} entry from the extracellular fluid. Shortly thereafter, contraction occurs.

The spontaneous depolarization-repolarization events occur in a regular and continuous manner in cardiac muscle, a property referred to as **rhythmicity.**

In the following experiment, you will observe these properties of cardiac muscle in a computer simulation. Additionally, your instructor may demonstrate this procedure using a real frog.

Nervous Stimulation of the Heart

Both the parasympathetic and sympathetic nervous systems innervate the heart. Stimulation of the sympathetic nervous system increases the rate and force of contraction of the heart. Stimulation of the parasympathetic nervous system (vagal nerves) decreases the depolarization rhythm of the sinoatrial node and slows transmission of excitation through the atrioventricular node. If vagal stimulation is excessive, the heart will stop beating. After a short time, the ventricles will begin to beat again. This is referred to as **vagal escape** and may be the result of sympathetic reflexes or initiation of a rhythm by the Purkinje fibers.

Baseline Frog Heart Activity

The heart's effectiveness as a pump is dependent both on intrinsic (within the heart) and extrinsic (external to the heart) controls. In this experiment, you will investigate some of these factors.

The nodal system, in which the "pacemaker" imposes its depolarization rate on the rest of the heart, is one intrinsic factor that influences the heart's pumping action. If its impulses fail to reach the ventricles (as in heart block), the ventricles continue to beat but at their own inherent rate, which is much slower than that usually imposed on them. Although heart contraction does not depend on nerve impulses, its rate can be modified by extrinsic impulses reaching it through the autonomic nerves. Additionally, cardiac activity is modified by various chemicals, hormones, ions, and metabolites. The effects of several of these chemical factors are examined in the next experimental series, Activities 4–9.

The frog heart has two atria and a single, incompletely divided ventricle. The pacemaker is located in the sinus venosus, an enlarged region between the venae cavae and the right atrium. The SA node of mammals may have evolved from the sinus venosus.

Choose **Frog Cardiovascular Physiology** from the main menu. The opening screen will appear in a few seconds (Figure 6.1). When the program starts, you will see a tracing of the frog's heartbeat on the *oscilloscope display* in the upper right part of the screen. Because the simulation automatically adjusts itself to your computer's speed, you may not see the heart tracing appear in real-time. If you want to increase the speed of the tracing (at the expense of tracing quality), click the **Tools** menu, choose **Modify Display,** and then select **Increase Speed.**

The oscilloscope display shows the ventricular contraction rate in the Heart Rate window. The *heart activity window* to the right of the Heart Rate display provides the following messages:

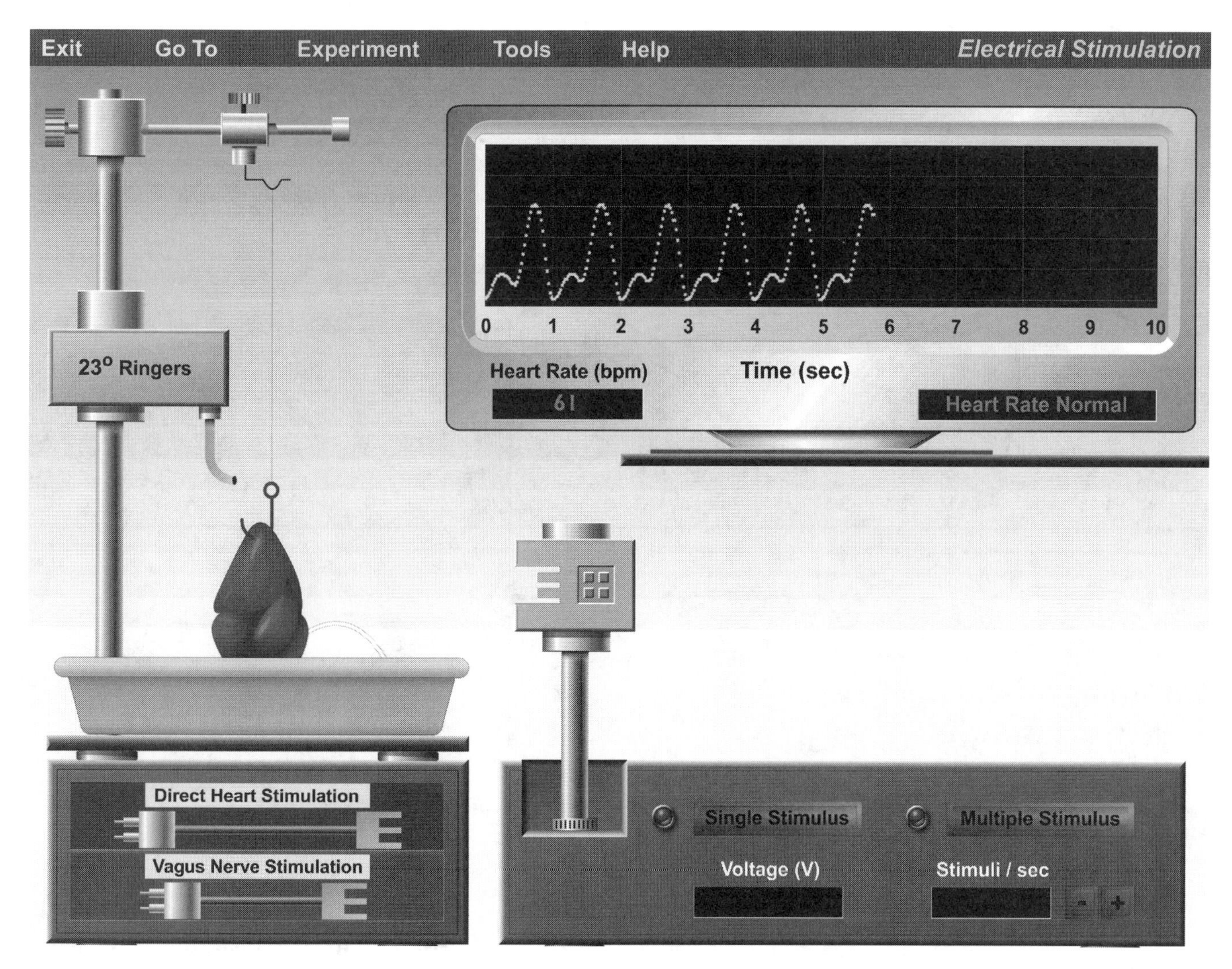

Figure 6.1 Opening screen of the Electrical Stimulation experiment.

- Heart Rate Normal—displayed when the heart is beating under resting conditions.
- Heart Rate Changing—displayed when the heart rate is increasing or decreasing.
- Heart Rate Stable—displayed when the heart rate is steady, but higher or lower than normal. For example, if you applied a chemical that increased heart rate to a stable but higher-than-normal rate, you would see this message.

The *electrical stimulator* is below the oscilloscope display. In the experiment, clicking **Single Stimulus** delivers a single electrical shock to the frog heart. Clicking **Multiple Stimulus** delivers repeated electrical shocks at the rate indicated in the Stimuli/sec window just below the Multiple Stimulus button. When the **Multiple Stimulus** button is clicked, it changes to a **Stop Stimulus** button that allows you to stop electrical stimulation as desired. Clicking the (+) or (−) buttons next to the Stimuli/sec window adjusts the stimulus rate. The voltage delivered when Single Stimulus or Multiple Stimulus is clicked is displayed in the Voltage window. The simulation automatically adjusts the voltage for the experiment. The postlike apparatus extending upward from the electrical stimulator is the *electrode holder* into which you will drag-and-drop electrodes from the supply cabinet in the bottom left corner of the screen.

The left side of the screen contains the apparatus that sustains the frog heart. The heart has been lifted away from the body of the frog by a hook passed through the apex of the heart. Although the frog cannot be seen because it is in the dissection tray, its heart has not been removed from its circulatory system. A thin string connects the hook in the heart to the force transducer at the top of the support bracket. As the heart contracts, the string exerts tension on the force transducer that converts the contraction into the oscilloscope tracing. The slender white strand extending from the heart toward the right side of the dissection tray is the vagus nerve. In the simulation, room-temperature (23°C) frog Ringer's solution continuously drips onto the heart to keep it moist and responsive so that a regular heart beat is maintained.

The two electrodes you will use during the experiment are located in the supply cabinet beneath the dissection tray. The Direct Heart Stimulation electrode is used to stimulate the ventricular muscle directly. The Vagus Nerve Stimulation electrode is used to stimulate the vagus nerve. To position either electrode, click and drag the electrode to the two-pronged plug in the electrode holder and then release the mouse button.

Activity 1:
Recording Baseline Frog Heart Activity

1. Before beginning to stimulate the frog heart experimentally, watch several heartbeats. Be sure you can distinguish atrial and ventricular contraction (Figure 6.2a).

2. Record the number of ventricular contractions per minute displayed in the Heart Rate window under the oscilloscope.

_______________ bpm (beats per minute) ■

Activity 2:
Investigating the Refractory Period of Cardiac Muscle

In Exercise 2 you saw that repeated rapid stimuli could cause skeletal muscle to remain in a contracted state. In other words, the muscle could be tetanized. This was possible because of the relatively short refractory period of skeletal muscle. In this experiment you will investigate the refractory period of cardiac muscle and its response to stimulation.

1. Click and hold the mouse button on the Direct Heart Stimulation electrode and drag it to the electrode holder.

2. Release the mouse button to lock the electrode in place. The electrode will touch the ventricular muscle tissue.

3. Deliver single shocks by clicking **Single Stimulus** at each of the following times. You may need to practice to acquire the correct technique.

- Near the beginning of ventricular contraction
- At the peak of ventricular contraction
- During the relaxation part of the cycle

Watch for **extrasystoles,** which are extra beats that show up riding on the ventricular contraction peak. Also note the compensatory pause, which allows the heart to get back on schedule after an extrasystole (Figure 6.2b).

During which portion of the cardiac cycle was it possible to induce an extrasystole?

____

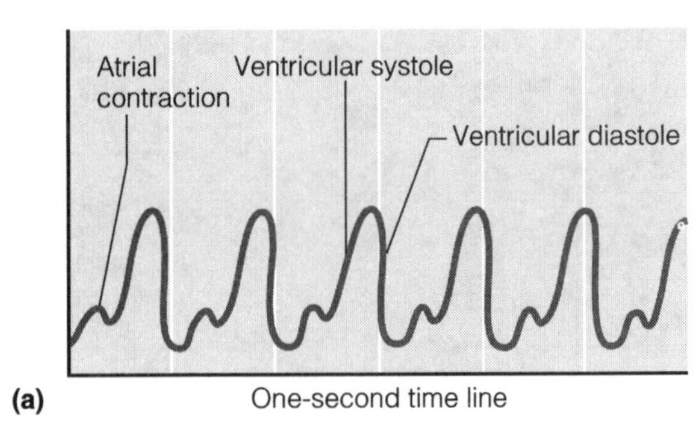

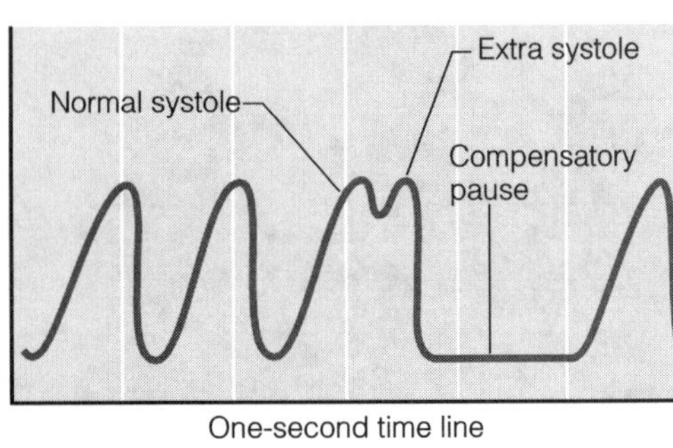

Figure 6.2 Recording of contractile activity of a frog heart. (a) Normal heartbeat. **(b)** Induction of an extrasystole.

4. Attempt to tetanize the heart by clicking **Multiple Stimulus.** Electrical shocks will be delivered to the muscle at a rate of 20 stimuli/sec. What is the result?

___

___

Considering the function of the heart, why is it important that the heart muscle cannot be tetanized?

___

___

5. Click **Stop Stimulus** to stop the electrical stimulation. ■

Activity 3:
Examining the Effect of Vagus Nerve Stimulation

The vagus nerve carries parasympathetic impulses to the heart, which modify heart activity.

1. Click the Direct Heart Stimulation electrode to return it to the supply cabinet.

2. Click and drag the Vagus Nerve Stimulation electrode to the electrode holder.

3. Release the mouse button to lock the electrode in place. The vagus nerve will automatically be draped over the electrode contacts.

4. Adjust the stimulator to 50 stimuli/sec by clicking the (+) or (−) buttons.

5. Click **Multiple Stimulus.** Allow the vagal stimulation to continue until the heart stops momentarily and then begins to beat again (vagal escape), and then click **Stop Stimulus.**

What is the effect of vagal stimulation on heart rate?

___

___

___

The phenomenon of vagal escape demonstrates that many factors are involved in heart regulation and that any deleterious factor (in this case, excessive vagal stimulation) will be overcome, if possible, by other physiological mechanisms such as activation of the sympathetic division of the autonomic nervous system (ANS). ■

Assessing Physical and Chemical Modifiers of Heart Rate

Now that you have observed normal frog heart activity, you will have an opportunity to investigate the effects of various modifying factors on heart activity. After removing the agent in each activity, allow the heart to return to its normal rate before continuing with the testing.

Choose **Modifiers of Heart Rate** from the **Experiment** menu. The opening screen will appear in a few seconds (Figure 6.3). The appearance and functionality of the *oscilloscope display* is the same as it was in the Electrical Stimulation experiment. The *solutions shelf* above the oscilloscope display contains the chemicals you'll use to modify heart rate in the experiment. You can choose the temperature of the Ringer's solution dispensed by clicking the appropriate button in the Ringer's dispenser at the left part of the screen. The doors to the supply cabinet are closed during this experiment because the electrical stimulator is not used.

When you click **Record Data** in the *data control unit* below the oscilloscope, your data is stored in the computer's memory and is displayed in the data grid at the bottom of the screen; data displayed include the solution used and the resulting heart rate. If you are not satisfied with a trial, you can click **Delete Line.** Click **Clear Table** if you wish to repeat the entire experiment.

Activity 4:
Assessing the Effect of Temperature

1. Click the **5°C Ringer's** button to bathe the frog heart in cold Ringer's solution. Watch the recording for a change in cardiac activity.

2. When the heart activity window displays the message Heart Rate Stable, click **Record Data** to retain your data in the data grid.

What change occurred with the cold (5°C) Ringer's solution?

___

3. Now click the **23°C Ringer's** button to flood the heart with fresh room-temperature Ringer's solution.

4. After you see the message Heart Rate Normal in the heart activity window, click the **32°C Ringer's** button.

5. When the heart activity window displays the message Heart Rate Stable, click **Record Data** to retain your data.

What change occurred with the warm (32°C) Ringer's solution?

___

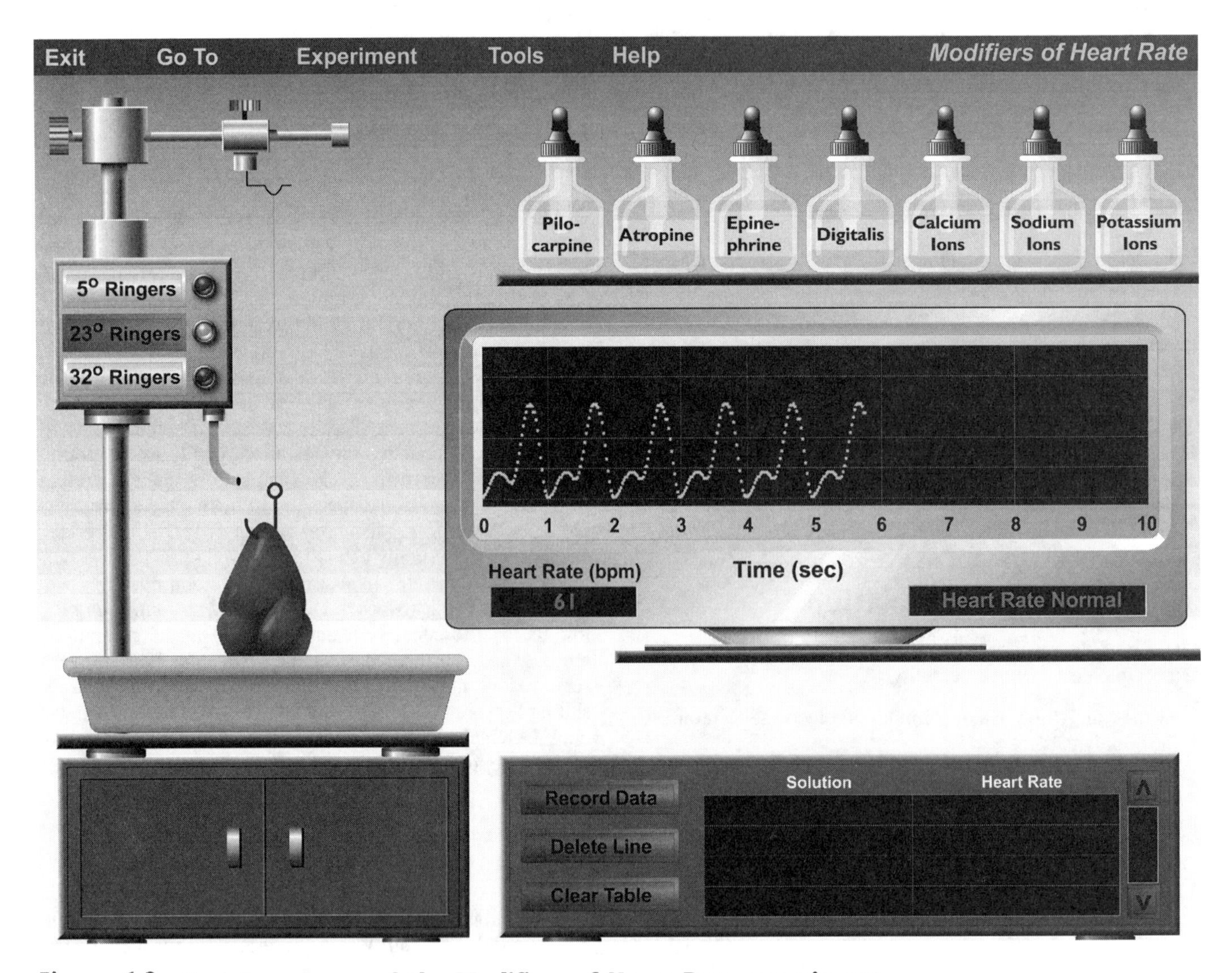

Figure 6.3 Opening screen of the Modifiers of Heart Rate experiment.

Record the heart rate at the two temperatures below.

_____________ bpm at 5°C; _____________ bpm at 32°C

What can you say about temperature and heart rate?

____

____

6. Click the **23°C Ringer's** button to flush the heart with fresh Ringer's solution. Watch the heart activity window for the message Heart Rate Normal before beginning the next test. ■

Activity 5:
Assessing the Effect of Pilocarpine

1. Click and hold the mouse on the Pilocarpine dropper cap.

2. Drag the dropper cap to a point about an inch above the heart and release the mouse.

3. Pilocarpine solution will be dispensed onto the heart and the dropper cap will automatically return to the Pilocarpine bottle.

4. Watch the heart activity window for the message Heart Rate Stable, indicating that the heart rate has stabilized under the effects of pilocarpine.

5. After the heart rate stabilizes, record the heart rate in the space provided below, and click **Record Data** to retain your data in the grid.

_____________ bpm

What happened when the heart was bathed in the pilocarpine solution?

____

6. Click the **23°C Ringer's** button to flush the heart with fresh Ringer's solution. Watch the heart activity window for the message Heart Rate Normal, an indication that the heart is ready for the next test. ■

Pilocarpine simulates the effect of parasympathetic nerve (hence, vagal) stimulation by enhancing acetylcholine release; such drugs are called parasympathomimetic drugs.

Activity 6: Assessing the Effect of Atropine

1. Drag-and-drop the Atropine dropper cap to a point about an inch above the heart.
2. Atropine solution will automatically drip onto the heart and the dropper cap will return to its position in the Atropine bottle.
3. Watch the heart activity window for the message Heart Rate Stable.
4. After the heart rate stabilizes, record the heart rate in the space below, and click **Record Data** to retain your data in the grid.

______________ bpm

What is the effect of atropine on the heart?

____

Atropine is a drug that blocks the effect of the neurotransmitter acetylcholine, liberated by the parasympathetic nerve endings. Do your results accurately reflect this effect of atropine?

____

Are pilocarpine and atropine agonists or antagonists in their effects on heart activity?

____

5. Click the **23°C Ringer's** button to flush the heart with fresh Ringer's solution. Watch the heart activity window for the message Heart Rate Normal before beginning the next test. ■

Activity 7: Assessing the Effect of Epinephrine

1. Drag-and-drop the Epinephrine dropper cap to a point about an inch above the heart.
2. Epinephrine solution will be dispensed onto the heart and the dropper cap will return to the Epinephrine bottle.
3. Watch the heart activity window for the message Heart Rate Stable.
4. After the heart rate stabilizes, record the heart rate in the space provided below, and click **Record Data** to retain your data in the grid.

______________ bpm

What happened when the heart was bathed in the epinephrine solution?

____

Which division of the autonomic nervous system does its effect imitate?

____

5. Click the **23°C Ringer's** button to flush the heart with fresh Ringer's solution. Watch the heart activity window for the message Heart Rate Normal, meaning that the heart is ready for the next test. ■

Activity 8: Assessing the Effect of Digitalis

1. Drag-and-drop the Digitalis dropper cap to a point about an inch above the heart.
2. Digitalis solution will automatically drip onto the heart and then the dropper will return to the Digitalis bottle.
3. Watch the heart activity window to the right of the Heart Rate window for the message Heart Rate Stable.
4. After the heart rate stabilizes, record the heart rate in the space provided below, and click **Record Data** to retain your data in the grid.

______________ bpm

What is the effect of digitalis on the heart?

____

5. Click the **23°C Ringer's** button to flush the heart with fresh Ringer's solution. Watch the heart activity window for the message Heart Rate Normal, then proceed to the next test. ■

Digitalis is a drug commonly prescribed for heart patients with congestive heart failure. It slows heart rate, providing more time for venous return and decreasing the workload on the weakened heart. These effects are thought to be due to inhibition of the Na^+-K^+ pump and enhancement of Ca^{2+} entry into myocardial fibers.

Activity 9:
Assessing the Effect of Various Ions

To test the effect of various ions on the heart, apply the desired solution using the following method.

1. Drag-and-drop the Calcium Ions dropper cap to a point about an inch above the heart.
2. Calcium ions will automatically be dripped onto the heart and the dropper cap will return to the Calcium Ions bottle.
3. Watch the heart activity window for the message Heart Rate Stable.
4. After the heart rate stabilizes, record the heart rate in the space provided below, and click **Record Data** to retain your data in the grid.
5. Click the **23°C Ringer's** button to flush the heart with fresh Ringer's solution. Watch the heart activity window for the message Heart Rate Normal, which means that the heart is ready for the next test.
6. Repeat steps 1 through 5 for Sodium Ions and then Potassium Ions.

Effect of Ca^{2+}:

Does the heart rate stabilize and remain stable?

____

Describe your observations of force and rhythm of the heartbeat.

____

____

Effect of Na^+:

Does the heart rate stabilize and remain stable?

____

Describe your observations of force and rhythm of the heartbeat.

____

____

Effect of K^+:

Does the heart rate stabilize and remain stable?

____

Describe your observations of force and rhythm of the heartbeat.

____

____

Potassium ion concentration is normally higher within cells than in the extracellular fluid. *Hyperkalemia* decreases the resting potential of plasma membranes, thus decreasing the force of heart contraction. In some cases, the conduction rate of the heart is so depressed that **ectopic pacemakers** (pacemakers appearing erratically and at abnormal sites in the heart muscle) appear in the ventricle, and fibrillation may occur.

Was there any evidence of premature beats in the recording of potassium ion effects?

____

Was arrhythmia produced with any of the ions tested?

______________ If so, which?

____ ■

Click **Tools** → **Print Data** to print your recorded data.

Histology Review Supplement

Turn to p. P-146 for a review of cardiovascular tissue.

xercise

7 Respiratory System Mechanics: Computer Simulation

Objectives

1. To define:

   ventilation, inspiration, expiration, forced expiration, tidal volume, vital capacity, expiratory reserve volume, inspiratory reserve volume, residual volume, surfactant, minute respiratory volume, forced expiratory volume, and *pneumothorax*
2. To describe the role of muscles and volume changes in the mechanics of breathing.
3. To understand that the lungs do not contain muscle and that respirations are therefore caused by external forces.
4. To explore the effect of changing airway resistance on breathing.
5. To study the effect of surfactant on lung function.
6. To examine the factors that cause lung collapse.
7. To understand the effects of hyperventilation, rebreathing, and breath holding on the CO_2 level in the blood.

The two phases of **pulmonary ventilation** or **breathing** are **inspiration,** during which air is taken into the lungs, and **expiration,** during which air is expelled from the lungs. Inspiration occurs as the external intercostal muscles and the diaphragm contract. The diaphragm, normally a dome-shaped muscle, flattens as it moves inferiorly while the external intercostal muscles between the ribs lift the rib cage. These cooperative actions increase the thoracic volume. Because the increase in thoracic volume causes a partial vacuum, air rushes into the lungs. During normal expiration, the inspiratory muscles relax, causing the diaphragm to rise and the chest wall to move inward. The thorax returns to its normal shape due to the elastic properties of the lung and thoracic wall. Like a deflating balloon, the pressure in the lungs rises, which forces air out of the lungs and airways. Although expiration is normally a passive process, abdominal wall muscles and the internal intercostal muscles can contract to force air from the lungs. Blowing up a balloon is an example where such **forced expiration** would occur.

Simulating Spirometry: Measuring Respiratory Volumes and Capacities

This computerized simulation allows you to investigate the basic mechanical function of the respiratory system as you determine lung volumes and capacities. The concepts you will learn by studying this simulated mechanical lung can then be applied to help you understand the operation of the human respiratory system.

Normal quiet breathing moves about 500 ml (0.5 liter) of air (the tidal volume) in and out of the lungs with each breath, but this amount can vary due to a person's size, sex, age, physical condition, and immediate respiratory needs. The terms used for the normal respiratory volumes are defined next. The values are for the normal adult male and are approximate.

Normal Respiratory Volumes

Tidal volume (TV): Amount of air inhaled or exhaled with each breath under resting conditions (500 ml)

Expiratory reserve volume (ERV): Amount of air that can be forcefully exhaled after a normal tidal volume exhalation (1200 ml)

Inspiratory reserve volume (IRV): Amount of air that can be forcefully inhaled after a normal tidal volume inhalation (3100 ml)

Residual volume (RV): Amount of air remaining in the lungs after complete exhalation (1200 ml)

Vital capacity (VC): Maximum amount of air that can be exhaled after a normal maximal inspiration (4800 ml)

$$VC = TV + IRV + ERV$$

Total lung capacity (TLC): Sum of vital capacity and residual volume

Pulmonary Function Tests

Forced vital capacity (FVC): Amount of air that can be expelled when the subject takes the deepest possible breath and exhales as completely and rapidly as possible

Forced expiratory volume (FEV_1): Measures the percentage of the vital capacity that is exhaled during 1 second of the FVC test (normally 75% to 85% of the vital capacity)

Choose **Respiratory System Mechanics** from the main menu. The opening screen for the Respiratory Volumes experiment will appear in a few seconds (Figure 7.1). The main features on the screen when the program starts are a pair of simulated lungs within a bell jar at the left side of the screen, an oscilloscope at the upper right part of the screen, a data display area beneath the oscilloscope, and a data control unit at the bottom of the screen.

The black rubber "diaphragm" sealing the bottom of the glass bell jar is attached to a rod in the pump just below the jar. The rod moves the rubber diaphragm up and down to change the pressure within the bell jar (comparable to the intrapleural pressure in the body). As the diaphragm moves inferiorly, the resulting volume increase creates a partial vacuum in the bell jar because of lowered pressure. This partial vacuum causes air to be sucked into the tube at the top of the bell jar and then into the simulated lungs.

Conversely, as the diaphragm moves up, the rising pressure within the bell jar forces air out of the lungs. The partition between the two lungs compartmentalizes the bell jar into right and left sides. The lungs are connected to an air-flow tube in which the diameter is adjustable by clicking the (+) and (−) buttons next to the Radius window in the equipment atop the bell jar. The volume of each breath that passes through the single air-flow tube above the bell jar is displayed

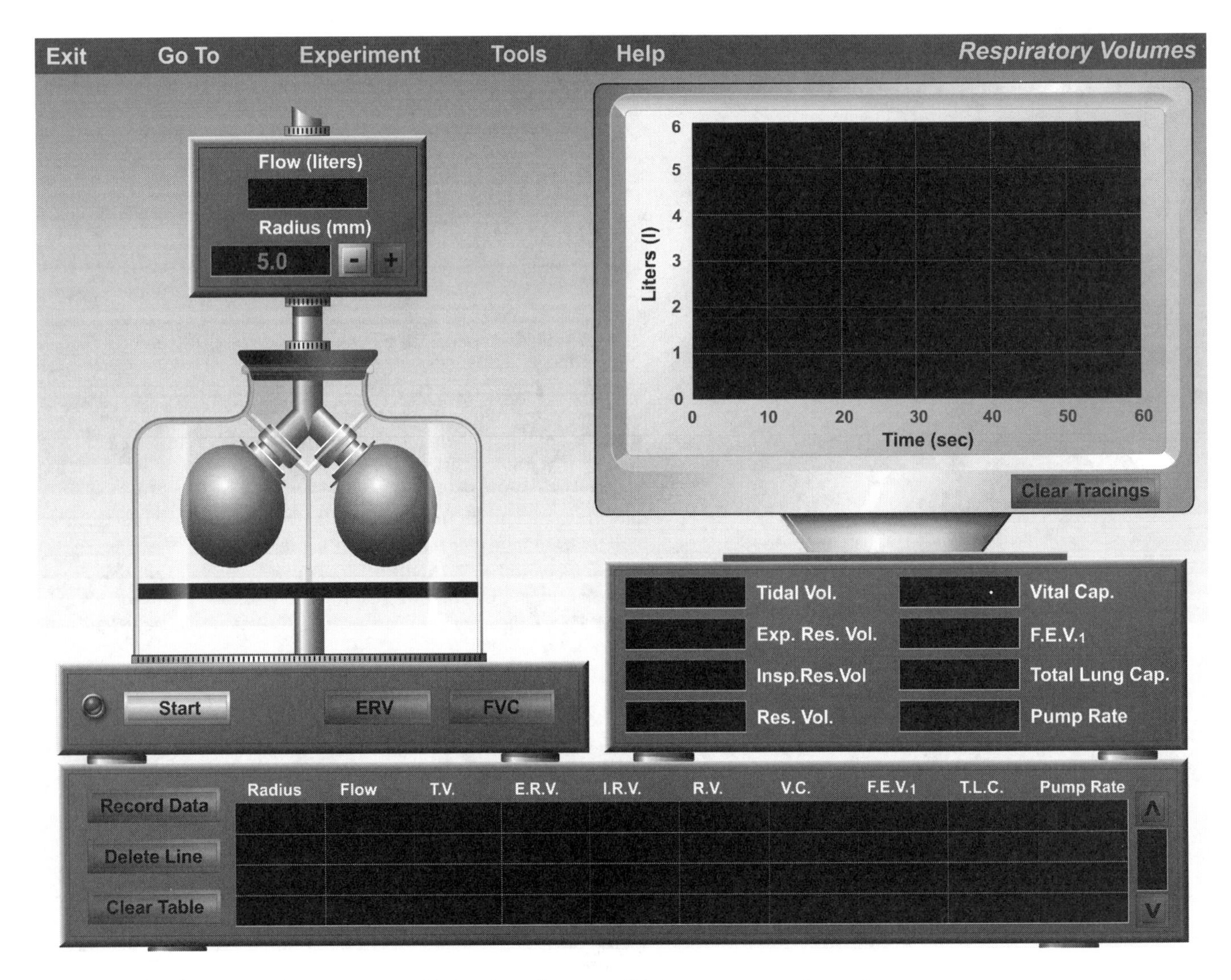

Figure 7.1 Opening screen of the Respiratory Volumes experiment.

in the Flow window. Clicking **Start** below the bell jar begins a trial run in which the simulated lungs will "breathe" in normal tidal volumes and the oscilloscope will display the tidal tracing. When **ERV** is clicked, the lungs will exhale maximally at the bottom of a tidal stroke and the expiratory reserve volume will be displayed in the Exp. Res. Vol. window below the oscilloscope. When **FVC** is clicked, the lungs will first inhale maximally and then exhale fully to demonstrate forced vital capacity. After ERV and FVC have been measured, the remaining lung values will be calculated and displayed in the small windows below the oscilloscope.

The data control equipment in the lower part of the screen records and displays data accumulated during the experiments. When you click **Record Data,** your data is recorded in the computer's memory and is displayed in the data grid. Data displayed in the data grid include the Radius, Flow, TV (tidal volume), ERV (expiratory reserve volume), IRV (inspiratory reserve volume), RV (residual volume), VC (vital capacity), FEV_1 (forced expiratory volume—1 second), TLC (total lung capacity), and Pump Rate. Clicking **Delete Line** allows you to discard the data for a single run; clicking **Clear Table** erases the entire experiment to allow you to start over.

If you need help identifying any piece of equipment, choose **Balloons On** from the Help menu and move the mouse pointer onto any piece of equipment visible on the computer's screen. As the pointer touches the object, a pop-up window appears to identify the equipment. To close the pop-up window, move the mouse pointer away from the equipment. Choose **Balloons Off** to turn off this help feature.

Activity 1: Measuring Respiratory Volumes

Your first experiment will establish the baseline respiratory values.

1. If the grid in the data control unit is not empty, click **Clear Table** to discard all previous data.

2. Adjust the radius of the airways to 5.00 mm by clicking the appropriate button next to the Radius window.

3. Click **Start** and allow the tracing to complete. Watch the simulated lungs begin to breathe as a result of the "contraction and relaxation" of the diaphragm. Simultaneously, the oscilloscope will display a tracing of the tidal volume for each breath. The Flow window atop the bell jar indicates the tidal volume for each breath, and the Tidal Vol. window below the oscilloscope shows the average tidal volume. The Pump Rate window displays the number of breaths per minute.

4. Click **Clear Tracings.**

5. Now click **Start** again. After a second or two, click **ERV,** wait 2 seconds and then click **FVC** to complete the measurement of respiratory volumes. The expiratory reserve volume, inspiratory reserve volume, and residual volume will be automatically calculated and displayed from the tests you have performed so far. Also, the equipment calculates and displays the total lung capacity.

6. Compute the **minute respiratory volume (MRV)** using the following formula (you can use the Calculator in the Tools menu):

$$\text{MRV} = \text{TV} \times \text{BPM (breaths per minute)}$$

MRV ____________ ml/min

7. Does expiratory reserve volume include tidal volume?

____

Explain your answer.

____

____

8. Now click **Record Data** to record the current experimental data in the data grid. Then click **Clear Tracings.**

9. If you want to print a tracing at any time, click **Tools** and then **Print Graph.** ■

Activity 2: Examining the Effect of Changing Airway Resistance on Respiratory Volumes

Lung diseases are often classified as obstructive or restrictive. With an obstructive problem, expiratory flow is affected, whereas a restrictive problem might indicate reduced inspiratory volume. Although they are not diagnostic, pulmonary function tests such as FEV_1 can help a clinician determine the difference between obstructive and restrictive problems. FEV_1 is the forced volume exhaled in one second. In obstructive disorders like chronic bronchitis and asthma, airway resistance is increased and FEV_1 will be low. Here you will explore the effect of changing the diameter of the airway on pulmonary function.

1. Do not clear the data table from the previous experiment.

2. Adjust the radius of the airways to 4.50 mm by clicking the appropriate button next to the Radius window.

3. Click **Start** to begin respirations.

4. Click **FVC.** As you saw in the previous test, the simulated lungs will inhale maximally and then exhale as forcefully as possible. FEV_1 will be displayed in the FEV_1 window below the oscilloscope.

5. When the lungs stop respiring, click **Record Data** to record the current data in the data grid.

6. Decrease the radius of the airways in 0.50-mm decrements and repeat steps 4 and 5 until the minimum radius (3.00 mm) is achieved. Be sure to click **Record Data** after each trial. Click **Clear Tracings** between trials. If you make an error and want to delete a single value, click the data line in the data grid and then click **Delete Line.**

7. A useful way to express FEV_1 is as a percentage of the forced vital capacity. Copy the FEV_1 and vital capacity values from the computer screen to the chart below and then calculate the FEV_1 (%) by dividing the FEV_1 volume by the vital capacity volume. Record the FEV_1 (%) in the chart. You can use the Calculator under the Tools menu.

What happened to the FEV_1 (%) as the radius of the airways was decreased?

____

Explain your answer.

____

____

Click **Tools → Print Data** to print your data. ■

Simulating Factors Affecting Respirations

This part of the computer simulation allows you to explore the action of surfactant on pulmonary function and the effect of changing the intrapleural pressure.

Choose **Factors Affecting Respirations** from the **Experiment** menu. The opening screen will appear in a few seconds (Figure 7.2). The basic features on the screen when the program starts are the same as in the Respiratory Volumes experiment screen. Additional equipment includes a surfactant dispenser atop the bell jar, and valves on each side of the bell jar. Each time **Surfactant** is clicked, a measured amount of surfactant is sprayed into the lungs. Clicking **Flush** washes surfactant from the lungs to prepare for another run.

Clicking the valve button (which currently reads **valve closed**) allows the pressure within that side of the bell jar to equalize with the atmospheric pressure. When **Reset** is clicked, the lungs are prepared for another run.

Data accumulated during a run are displayed in the windows below the oscilloscope. When you click **Record Data,** that data is recorded in the computer's memory and is displayed in the data grid. Data displayed in the data grid include the Radius, Pump Rate, the amount of Surfactant, Pressure Left (pressure in the left lung), Pressure Right (pressure in the right lung), Flow Left (air flow in the left lung), Flow Right (air flow in the right lung), and Total Flow. Clicking **Delete Line** allows you to discard data values for a single run, and clicking **Clear Table** erases the entire experiment to allow you to start over.

Activity 3: Examining the Effect of Surfactant

At any gas-liquid boundary, the molecules of the liquid are attracted more strongly to each other than they are to the air molecules. This unequal attraction produces tension at the liquid surface called surface tension. Because surface tension resists any force that tends to increase surface area, it acts to decrease the size of hollow spaces, such as the alveoli or microscopic air spaces within the lungs. If the film lining the air spaces in the lung were pure water, it would be very difficult, if not impossible, to inflate the lungs. However, the aqueous film covering the alveolar surfaces contains **surfactant,** a detergent-like lipoprotein that decreases surface tension by reducing the attraction of water molecules for each other. You will explore the action of surfactant in this experiment.

1. If the data grid is not empty, click **Clear Table** to discard all previous data values.
2. Adjust the airway radius to 5.00 mm by clicking the appropriate button next to the Radius window.
3. If necessary, click **Flush** to clear the simulated lungs of existing surfactant.
4. Click **Start** and allow a baseline run without added surfactant to complete.
5. When the run completes, click **Record Data.**
6. Now click **Surfactant** twice.
7. Click **Start** to begin the surfactant run.
8. When the lungs stop respiring, click **Record Data** to display the data in the grid.

FEV_1 as % of VC

Radius	FEV_1	Vital Capacity	FEV_1 (%)
5.00			
4.50			
4.00			
3.50			
3.00			

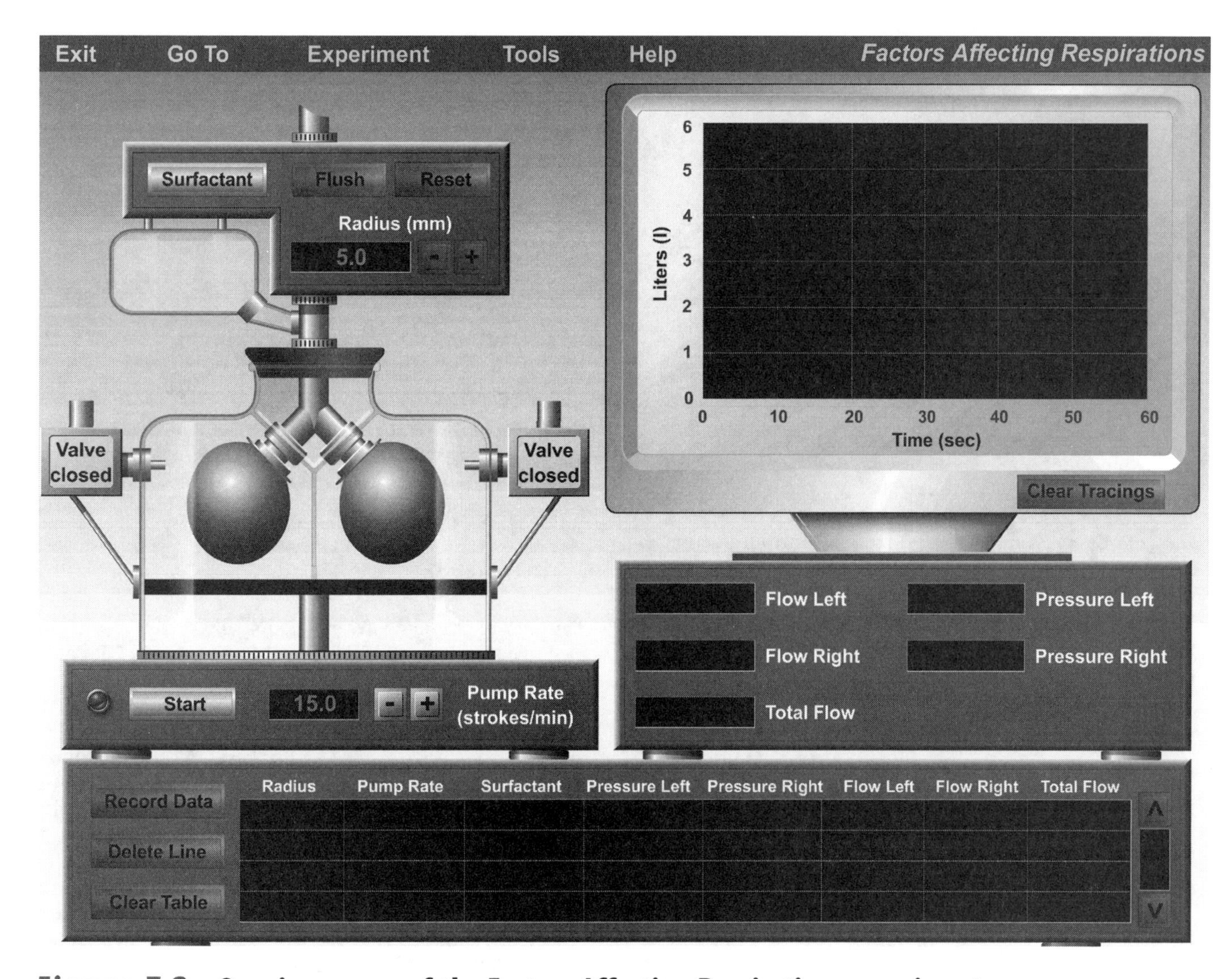

Figure 7.2 Opening screen of the Factors Affecting Respirations experiment.

How has the air flow changed compared to the baseline run?

____

____

Premature infants often have difficulty breathing. Explain why this might be so. (Use your text as needed.)

____

____ ■

Activity 4: Investigating Intrapleural Pressure

The pressure within the pleural cavity, **intrapleural pressure,** is less than the pressure within the alveoli. This negative pressure condition is caused by two forces, the tendency of the lung to recoil due to its elastic properties and the surface tension of the alveolar fluid. These two forces act to pull the lungs away from the thoracic wall, creating a partial vacuum in the pleural cavity. Because the pressure in the intrapleural space is lower than atmospheric pressure, any opening created in the thoracic wall equalizes the intrapleural pressure with the atmospheric pressure, allowing air to enter the pleural cavity, a condition called **pneumothorax.** Pneumothorax allows lung collapse, a condition called **atelectasis** (at″ĕ-lik′tah-sis).

In the simulated respiratory system on the computer's screen, the intrapleural space is the space between the wall of the bell jar and the outer wall of the lung it contains. The pressure difference between inspiration and expiration for the left lung and for the right lung is individually displayed in the Pressure Left and Pressure Right windows.

1. Do not discard your previous data.

2. Click **Clear Tracings** to clean up the screen and then click **Flush** to clear the lungs of surfactant from the previous run.

3. Adjust the radius of the airways to 5.00 mm by clicking the appropriate button next to the Radius window.

4. Click **Start** and allow one screen of respirations to complete. Notice the negative pressure condition displayed below the oscilloscope when the lungs inhale.

5. When the lungs stop respiring, click **Record Data** to display the data in the grid.

6. Now click the valve button (which currently reads valve closed) on the left side of the bell jar above the Start button to open the valve.

7. Click **Start** to begin the run.

8. When the run completes, click **Record Data** again.

What happened to the lung in the left side of the bell jar?

____

____

How did the pressure in the left lung differ from that in the right lung?

____

Explain your reasoning.

____

____

How did the total air flow in this trial compare with that in the previous trial in which the pleural cavities were intact?

____

____

What do you think would happen if the two lungs were in a single large cavity instead of separate cavities?

____

____

9. Now close the valve you opened earlier by clicking it and then click **Start** to begin a new trial.

10. When the run completes, click **Record Data** to display the data in the grid.

Did the deflated lung reinflate? _______________

Explain your answer.

____

____

11. Click the **Reset** button atop the bell jar. This action draws the air out of the intrapleural space and returns it to normal resting condition.

12. Click **Start** and allow the run to complete.

13. When the run completes, click **Record Data** to display the data in the grid.

Why did lung function in the deflated (left) lung return to normal after you clicked Reset?

____

____ ■

Click **Tools → Print Data** to print your data.

Simulating Variations in Breathing

This part of the computer simulation allows you to examine the effects of hyperventilation, rebreathing, and breath holding on CO_2 level in the blood.

Choose **Variations in Breathing** from the **Experiment** menu. The opening screen will appear in a few seconds (Figure 7.3). The basic features on the screen when the program starts are the same as in the lung volumes screen. The buttons beneath the oscilloscope control the various possible breathing patterns. Clicking **Rapid Breathing** causes the lungs to breathe faster than normal. A small bag automatically covers the airway tube when **Rebreathing** is clicked. Clicking **Breath Holding** causes the lungs to stop respiring. Click **Normal Breathing** at any time to resume normal tidal cycles. The window next to the Start button displays the breathing pattern being performed by the simulated lungs.

The windows below the oscilloscope display the PCO_2 (partial pressure of CO_2) of the air in the lungs, Maximum PCO_2, Minimum PCO_2, and Pump rate.

Data accumulated during a run are displayed in the windows below the oscilloscope. When you click **Record Data,** that data is recorded in the computer's memory and is displayed in the data grid. Data displayed in the data grid include the Condition, PCO_2, Max. PCO_2, Min. PCO_2, Pump Rate, Radius, and Total Flow. Clicking **Delete Line** allows you to discard data values for a single run, and clicking **Clear Table** erases the entire experiment to allow you to start over.

Activity 5:
Exploring Various Breathing Patterns

You will establish the baseline respiratory values in this first experiment.

1. If the grid in the data control unit is not empty, click **Clear Table** to discard all previous data.

2. Adjust the radius of the airways to 5.00 mm by clicking the appropriate button next to the Radius window. Now, read through steps 3–5 before attempting to execute them.

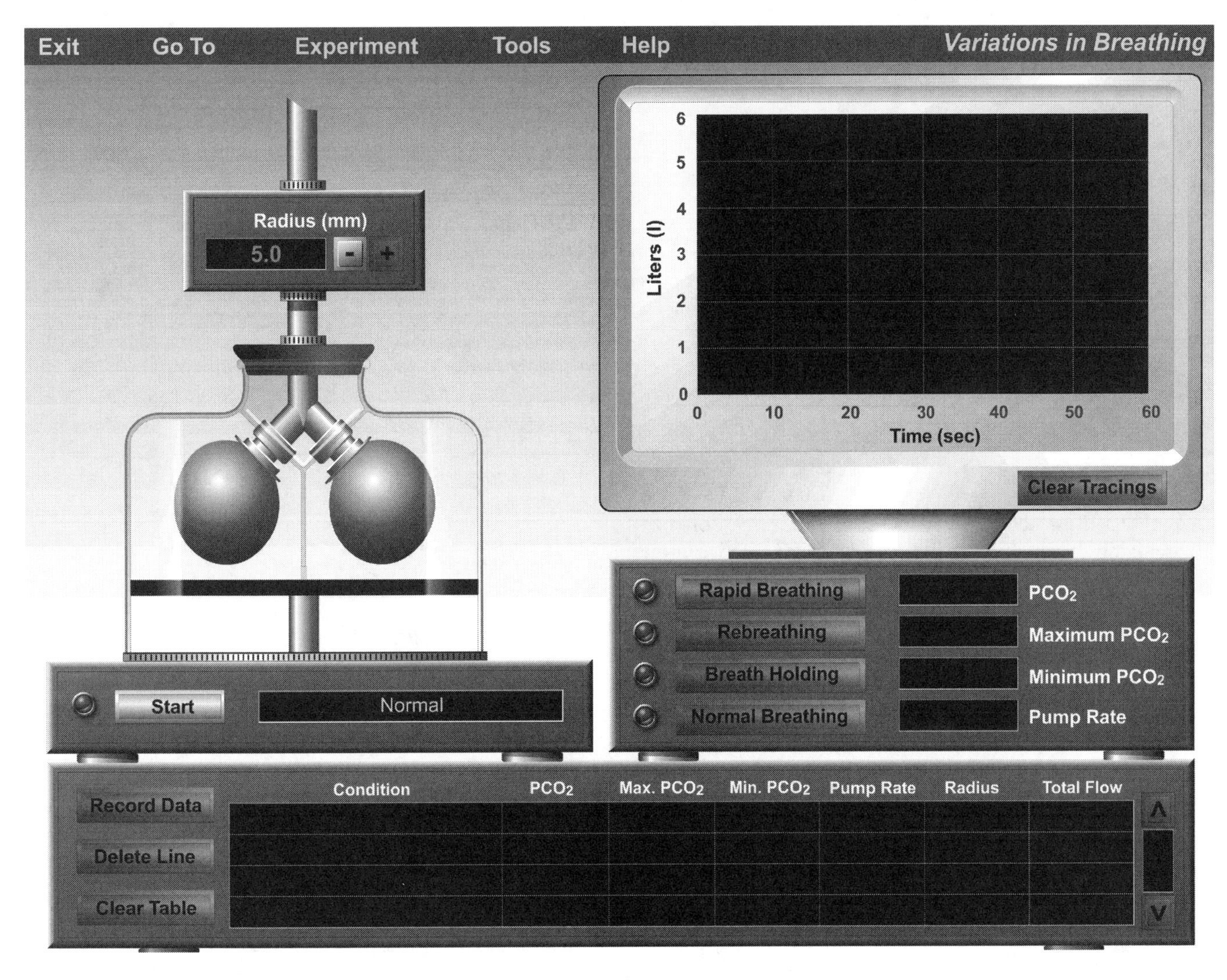

Figure 7.3 Opening screen of the Variations in Breathing experiment.

3. Click **Start** and notice that it changes to **Stop** to allow you to stop the respiration. Watch the simulated lungs begin to breathe as a result of the external mechanical forces supplied by the pump below the bell jar. Simultaneously, the oscilloscope will display a tracing of the tidal volume for each breath.

4. After 2 seconds, click the **Rapid Breathing** button and watch the PCO_2 displays. The breathing pattern will change to short, rapid breaths. The PCO_2 of the air in the lungs will be displayed in the small window to the right of the **Rapid Breathing** button.

5. Watch the oscilloscope display and the PCO_2 window, and click **Stop** before the tracing reaches the end of the screen.

What happens to PCO_2 during rapid breathing? Explain your answer.

____

____

6. Click **Record Data.**

7. Now click **Clear Tracings** to prepare for the next run.

Rebreathing

When **Rebreathing** is clicked, a small bag will appear over the end of the air tube to allow the air within the lungs to be repeatedly inspired and expired.

1. Click **Start,** wait 2 seconds, and then click **Rebreathing.**

2. Watch the breathing pattern on the oscilloscope and notice the PCO_2 during the course of the run. Click **Stop** when the tracing reaches the right edge of the oscilloscope.

What happens to PCO_2 during the entire time of the rebreathing activity?

____

____

Did the depth of the breathing pattern change during rebreathing? (Carefully examine the tracing for rate and depth changes; the changes can be subtle.) Explain.

____

____

____

____

3. Click **Record Data** and then click **Clear Tracings** to prepare for the next run.

Breath Holding

Breath holding can be considered an extreme form of rebreathing in which there is no gas exchange between the outside atmosphere and the air within the lungs.

1. Click **Start,** wait a second or two, and then click **Breath Holding.**

2. Let the breath-holding activity continue for about 5 seconds and then click **Normal Breathing.**

3. Click **Stop** when the tracing reaches the right edge of the oscilloscope.

What happened to the PCO_2 during breath holding?

____

What happened to the breathing pattern when normal respirations resume?

____

____

4. Click **Record Data.**

Click **Tools** → **Print Data** to print your recorded data. ■

Histology Review Supplement

Turn to p. P-147 for a review of respiratory tissue.

Chemical and Physical Processes of Digestion: Computer Simulation

Exercise

8

Objectives

1. To list the digestive system enzymes involved in the digestion of proteins, fats, and carbohydrates; to state their site of origin; and to summarize the environmental conditions promoting their optimal functioning.
2. To recognize the variation between different types of enzyme assays.
3. To name the end products of digestion of proteins, fats, and carbohydrates.
4. To perform the appropriate chemical tests to determine if digestion of a particular food has occurred.
5. To cite the function(s) of bile in the digestive process.
6. To discuss the possible role of temperature and pH in the regulation of enzyme activity.
7. To define *enzyme, catalyst, control, substrate,* and *hydrolase.*
8. To explain why swallowing is both a voluntary and a reflex activity.
9. To discuss the role of the tongue, larynx, and gastroesophageal sphincter in swallowing.
10. To compare and contrast segmentation and peristalsis as mechanisms of propulsion.

The digestive system is a physiological marvel, composed of finely orchestrated chemical and physical activities. The food we ingest must be broken down to its molecular form for us to get the nutrients we need, and digestion involves a complex sequence of mechanical and chemical processes designed to achieve this goal as efficiently as possible. As food passes through the gastrointestinal tract, it is progressively broken down by the mechanical action of smooth muscle and the chemical action of enzymes until most nutrients have been extracted and absorbed into the blood.

Chemical Digestion of Foodstuffs: Enzymatic Action

Nutrients can only be absorbed when broken down into their monomer form, so food digestion is a prerequisite to food absorption. **Enzymes** are large protein molecules produced by body cells. They are biological **catalysts** that increase the rate of a chemical reaction without becoming part of the product. The digestive enzymes are hydrolytic enzymes, or **hydrolases,** which break down organic food molecules or **substrates** by adding water to the molecular bonds, thus cleaving the bonds between the subunits or monomers.

A hydrolytic enzyme is highly specific in its action. Each enzyme hydrolyzes one or, at most, a small group of substrate molecules, and specific environmental conditions are necessary for an enzyme to function optimally. For example, temperature and pH have a large effect on the degree of enzymatic hydrolysis, and each enzyme has its preferred environment.

Because digestive enzymes actually function outside the body cells in the digestive tract lumen, their hydrolytic activity can also be studied in a test tube. Such *in vitro* studies provide a convenient laboratory environment for investigating the effect of various factors on enzymatic activity.

Starch Digestion by Salivary Amylase

In this experiment you will investigate the hydrolysis of starch to maltose by salivary amylase, the enzyme produced by the salivary glands and secreted into the mouth. For you to be able to detect whether or not enzymatic action has occurred, you need to be able to identify the presence of these substances to determine to what extent hydrolysis has occurred. Thus, **controls** must be prepared to provide a known standard against which comparisons can be made. The controls will vary for each experiment, and will be discussed in each enzyme section in this exercise.

Starch decreases and sugar increases as digestion proceeds according to the following equation:

$$\text{Starch} + \text{water} \xrightarrow{\text{amylase}} \text{X maltose}$$

Because the chemical changes that occur as starch is digested to maltose cannot be seen by the naked eye, you need to conduct an *enzyme assay,* the chemical method of detecting the presence of digested substances. You will perform two enzyme assays on each sample. The IKI assay detects the presence of starch and the Benedict's assay tests for the presence of maltose, which is the digestion product of starch. Normally a caramel-colored solution, IKI turns blue-black in the presence of starch. Benedict's reagent is a bright blue solution that changes to green to orange to reddish-brown with increasing amounts of maltose. It is important to understand that enzyme assays only indicate the presence or absence of substances. It is up to you to analyze the results of the experiments to decide if enzymatic hydrolysis has occurred.

Choose **Chemical and Physical Processes of Digestion** from the main menu. The opening screen will appear in a few seconds (Figure 8.1). The *solutions shelf* in the upper right part of the screen contains the substances to be used in the experiment. The *incubation unit* beneath the solutions shelf contains a rack of test tube holders and the apparatus needed to run the experiments. Test tubes from the *test tube washer* on the left part of the screen are loaded into the rack in the incubation unit by clicking and holding the mouse button on the first tube, and then releasing (dragging-and-dropping) it into any position in the rack. The substances in the dropper bottles on the solutions shelf are dispensed by dragging-and-dropping the dropper cap to a position over any test tube in

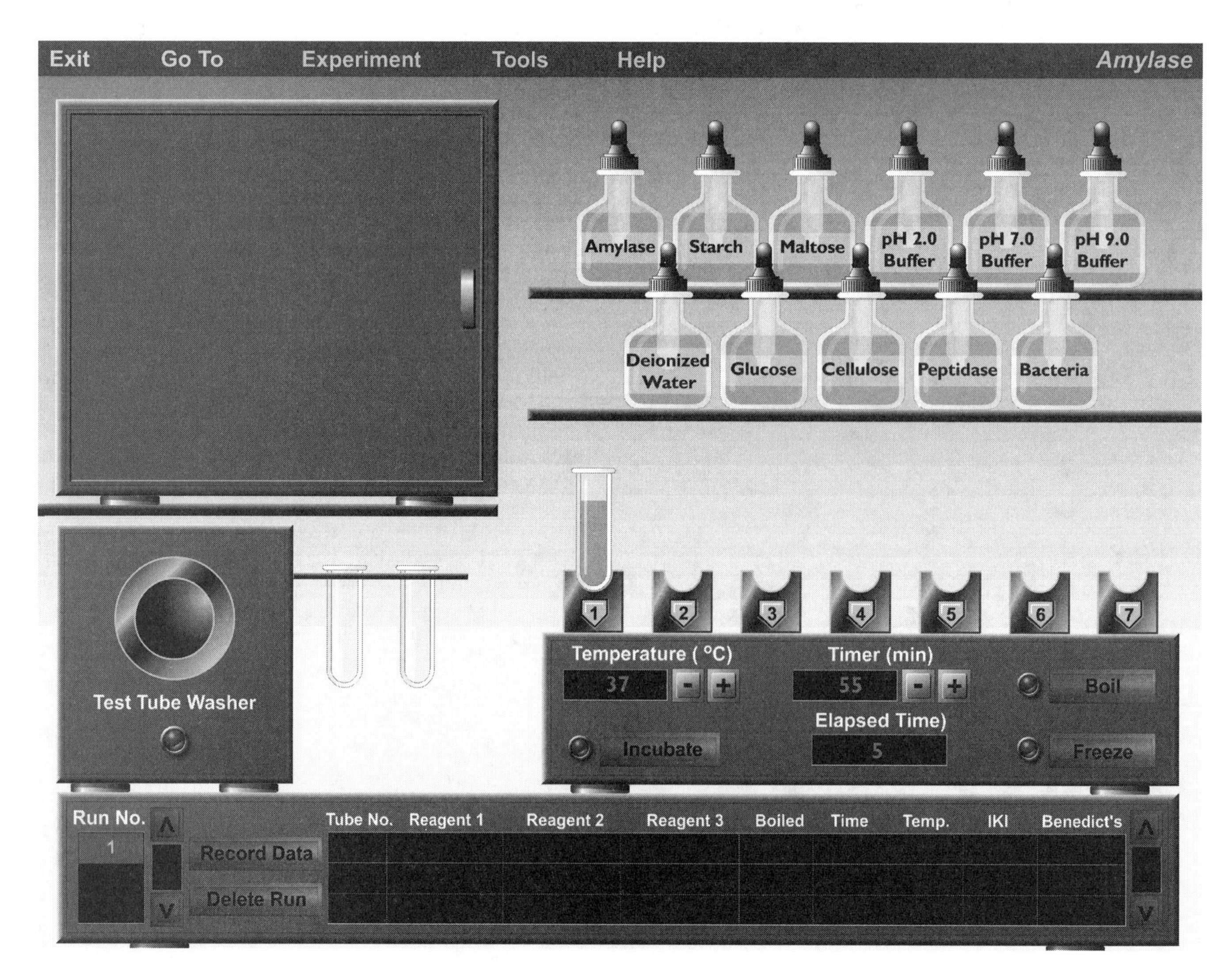

Figure 8.1 Opening screen of the Amylase experiment.

the rack and then releasing it. During each dispensing event, five drops of solution drip into the test tube; then the dropper cap automatically returns to its position in the bottle.

Each test tube holder in the incubation unit not only supports but also allows you to boil the contents of a single test tube. Clicking the numbered button at the base of a test tube holder causes that single tube to descend into the incubation unit. To boil the contents of all tubes inside the incubation unit, click **Boil.** After they have been boiled, the tubes automatically rise. You can adjust the incubation temperature for the experiment by clicking the (+) or (−) buttons next to the Temperature window. Set the incubation time by clicking the (+) or (−) buttons next to the Timer window. Clicking the **Incubate** button starts the timer and causes the entire rack of tube holders to descend into the incubation unit where the tubes will be incubated at the temperature and the time indicated. While incubating, the tubes are automatically agitated to ensure that their contents are well mixed. During the experiment, elapsed time is displayed in the Elapsed Time window.

The cabinet doors in the *assay cabinet* above the test tube washer are closed at the beginning of the experiment, but they automatically open when the set time for incubation has elapsed. The assay cabinet contains the reagents and glassware needed to assay your experimental samples.

When you click the **Record Data** button in the *data control unit* at the bottom of the screen, your data is recorded in the computer's memory and displayed in the data grid at the bottom of the screen. Data displayed in the grid include the tube number, the three substances dispensed into each tube, the time and incubation temperature, and (+) or (−) marks indicating enzyme assay results and whether or not a sample was boiled. If you are not satisfied with a single run, you can click **Delete Run** to erase an experiment.

Once an experimental run is completed and you have recorded your data, discard the test tubes to prepare for a new run by dragging the used tubes to the large opening in the test tube washer. The test tubes will automatically be prepared for the next experiment.

Activity 1:
Assessing Animal Starch Digestion by Salivary Amylase

Incubation

1. Individually drag seven test tubes to the test tube holders in the incubation unit.

2. Prepare tubes 1 through 7 with the substances indicated in the chart below using the following approach.

- Click and hold the mouse button on the dropper cap of the desired substance on the solutions shelf.
- While still holding the mouse button down, drag the dropper cap to the top of the desired test tube.
- Release the mouse button to dispense the substance. The dropper cap automatically returns to its bottle.

Note that the starch we are using here is animal starch (glycogen).

3. When all tubes are prepared, click the number **(1)** under the first test tube. The tube will descend into the incubation unit. All other tubes should remain in the raised position.

4. Click **Boil** to boil the number 1 tube. After boiling for a few moments, the tube will automatically rise.

5. Now adjust the incubation temperature to 37°C and the timer to 60 min (compressed time) by clicking the **(+)** or **(−)** buttons.

6. Click **Incubate** to start the run. The incubation unit will gently agitate the test tube rack, evenly mixing the contents of all test tubes throughout the incubation. Notice that the computer compresses the 60-minute time period into 60 seconds of real time, so what would be a 60-minute incubation in real life will take only 60 seconds in the simulation. When the incubation time elapses, the test tube rack will automatically rise, and the doors to the assay cabinet will open.

Salivary Amylase Digestion of Animal Starch

Tube no.	1	2	3	4	5	6	7
Additives	Amylase Starch pH 7.0 buffer	Amylase Starch pH 7.0 buffer	Amylase Deionized water pH 7.0 buffer	Deionized water Starch pH 7.0 buffer	Deionized water Maltose pH 7.0 buffer	Amylase Starch pH 2.0 buffer	Amylase Starch pH 9.0 buffer
Incubation condition	Boil first, then incubate at 37°C	37°C	37°C	37°C	37°C	37°C	37°C
Benedict's test							
IKI test							

Assays

After the assay cabinet doors open, notice the two reagents in the assay cabinet. IKI tests for the presence of starch and Benedict's detects the presence of glucose or maltose, the digestion product of starch. Below the reagents are seven small assay tubes into which you will dispense a small amount of test solution from the incubated samples in the incubation unit, plus a drop of IKI.

1. Click and hold the mouse on the first tube in the incubation unit. Notice that the mouse pointer is now a miniature test tube tilted to the left.

2. While still holding the mouse button down, move the mouse pointer to the first small assay tube on the left side of the assay cabinet. Release the mouse button. Watch the first test tube automatically decant approximately half of its contents into the first assay tube on the left.

3. Repeat steps 1 and 2 for the remaining tubes in the incubation unit, moving to a fresh assay tube each time.

4. Next, click and hold the mouse on the IKI dropper cap and drag it to the first assay tube. Release the mouse button to dispense a drop of IKI into the first assay tube on the left. You will see IKI drip into the tube, which may cause a color change in the solution. A blue-black color indicates a **positive starch test.** If starch is not present, the mixture will look like diluted IKI, a **negative starch test.** Intermediate starch amounts result in a pale gray color.

5. Now dispense IKI into the remaining assay tubes. Record your results (+ for positive, − for negative) in the chart.

6. Dispense Benedict's reagent into the remaining mixture in each tube in the incubation unit by dragging-and-dropping the Benedict's dropper cap to the top of each test tube.

7. After Benedict's reagent has been delivered to each tube in the incubation unit, click **Boil.** The entire tube rack will descend into the incubation unit and automatically boil the tube contents for a few moments.

8. When the rack of tubes rises, inspect the tubes for color change. A green-to-reddish color indicates that maltose is present; this is a **positive sugar test.** An orange-colored sample contains more maltose than a green sample. A reddish-brown color indicates even more maltose. A negative sugar test is indicated by no color change from the original bright blue. Record your results in the chart.

9. Click **Record Data** to display your results in the grid and retain your data in the computer's memory for later analysis. To repeat the experiment, drag all test tubes to the test tube washer and start again.

10. Answer the following questions, referring to the chart (or the data grid in the simulation) as necessary. Hint: closely examine the IKI and Benedict's results for each tube.

What do tubes 2, 6, and 7 reveal about pH and amylase activity?

____

____

Which pH buffer allowed the highest amylase activity?

____

Which tube indicates that the amylase did not contain maltose? ______________________________

Which tubes indicate that the deionized water did not contain starch or maltose? ______________________________

If we left out control tubes 3, 4, or 5, what objections could be raised to the statement: "Amylase digests starch to maltose"? (Hint: think about the purity of the chemical solutions.)

____

____

Would the amylase present in saliva be active in the stomach? Explain your answer.

____

____

What effect does boiling have on enzyme activity?

____

______________________________ ■

Activity 2:
Assessing Cellulose Digestion

If any test tubes are still in the incubator, click and drag them to the test tube washer before beginning this activity.

In the previous activity, we learned that salivary amylase can digest animal starch. In this activity, we will test to see whether amylase digests **cellulose,** a component of the cell walls of plants. We will also investigate whether bacteria (such as that found in the stomachs of cows and other ruminants) will digest cellulose.

Incubation

1. Individually drag seven test tubes to the test tube holders in the incubation unit.

2. Prepare tubes 1 through 7 with the substances indicated in the chart on the next page using the following approach:

- Click and hold the mouse button on the dropper cap of the desired substance on the solutions shelf.

- While holding the mouse button down, drag the dropper cap to the top of the desired test tube.

- Release the mouse button to dispense the substance. The dropper cap automatically returns to its bottle.

Note that the Starch bottle contains animal starch (glycogen).

3. When all tubes are prepared, click the number **(1)** under the first test tube. The tube will descend into the incubation unit.

4. Click **Freeze.** The tube's contents will be subjected to a temperature of −25°C. The tube will then automatically rise, with the contents of the tube frozen.

5. Adjust the incubation temperature to 37°C and the timer to 60 minutes by clicking the (+) and (−) buttons.

6. Click **Incubate** to start the run. The incubation unit will gently agitate the tubes as they incubate to thoroughly mix the tubes' contents. At the end of the incubation period, the tubes will ascend to their original positions on top of the incubator, and the doors to the assay cabinet will open.

Assays

When the assay cabinet opens, notice the two reagents in the cabinet. They are the same ones as in the previous activity: IKI will test for the presence of starch while Benedict's solution will test for the presence of glucose or maltose. On the floor of the cabinet are seven small tubes that you will use to test the results of your experiment. The procedure will be identical to the one from the previous activity:

1. Click and hold the mouse on the first tube in the incubation unit. Notice that the mouse pointer is now a miniature test tube tilted to the left.

2. While still holding the mouse button down, move the mouse pointer to the first small assay tube on the left side of the assay cabinet. Release the mouse button.

3. Repeat steps 1 and 2 for the remaining tubes in the incubation unit, moving to a fresh assay tube each time.

4. Next, click and hold the mouse on the IKI dropper cap and drag it to the first assay tube. Release the mouse button to dispense a drop of IKI into the first assay tube on the left. You will see IKI drip into the tube, which may cause a color change in the solution. A blue-black color indicates the presence of starch. If there is only a small amount of starch, you may see a pale gray color. If starch is not present, the mixture will look like diluted IKI.

5. Place IKI into the remaining assay tubes and note the color of each tube. Record your results in the chart.

6. Dispense Benedict's reagent into the remaining mixture in each tube in the incubation unit by dragging-and-dropping the Benedict's dropper cap to the top of each test tube.

7. After Benedict's reagent has been delivered to each tube in the incubation unit, click **Boil.** The entire tube rack will descend into the incubation unit and automatically boil the tube contents for a few moments.

8. When the rack of tubes rises, inspect the tubes for color change. A green to reddish color indicates glucose or maltose is present for a positive sugar test. An orange sample indicates more sugar than the green color, while a reddish-brown indicates the highest amounts of sugar. If there has been no color change from the original blue, no glucose or maltose is present in the tube. Record your results in the chart.

9. Click **Record Data** to display your results in the grid and retain your data in the computer's memory for later analysis. To repeat the experiment, drag all test tubes to the test tube washer and start again.

10. Answer the following questions, referring to the chart.

Which tubes showed a positive test for the IKI reagent?

____

Which tubes showed a positive test for the Benedict's reagent? _______________________________________

What was the effect of freezing tube 1?

____

Enzyme Digestion of Animal Starch and Cellulose

Tube no.	1	2	3	4	5	6	7
Additives	Amylase Starch pH 7.0 buffer	Amylase Starch pH 7.0 buffer	Amylase Glucose pH 7.0 buffer	Amylase Cellulose pH 7.0 buffer	Amylase Cellulose Deionized water	Peptidase Starch pH 7.0 buffer	Bacteria Cellulose pH 7.0 buffer
Incubation condition	Freeze first, then incubate at 37°C	37°C	37°C	37°C	37°C	37°C	37°C
Benedict's test							
IKI test							

How does the effect of freezing differ from the effect of boil-ing? ______________________________

____

What was the effect of amylase on glucose in tube 3? Can you offer an explanation for this effect? ______________

____

____

What was the effect of amylase on cellulose in tube 4?

____

____

Popcorn and celery are nearly pure plant starch or cellulose. What can you conclude about the digestion of cellulose, judging from the results of test tubes 4, 5, and 7?

____

____

What was the effect of the different enzyme, peptidase, used in tube 6? Explain your answer, based on what you know about peptidase.

____

______________________________ ■

Click **Tools → Print Data** to print your data.

Protein Digestion by Pepsin

The chief cells of the stomach glands produce pepsin, a protein-digesting enzyme. Pepsin hydrolyzes proteins to small fragments (proteoses, peptones, and peptides). In this experi-

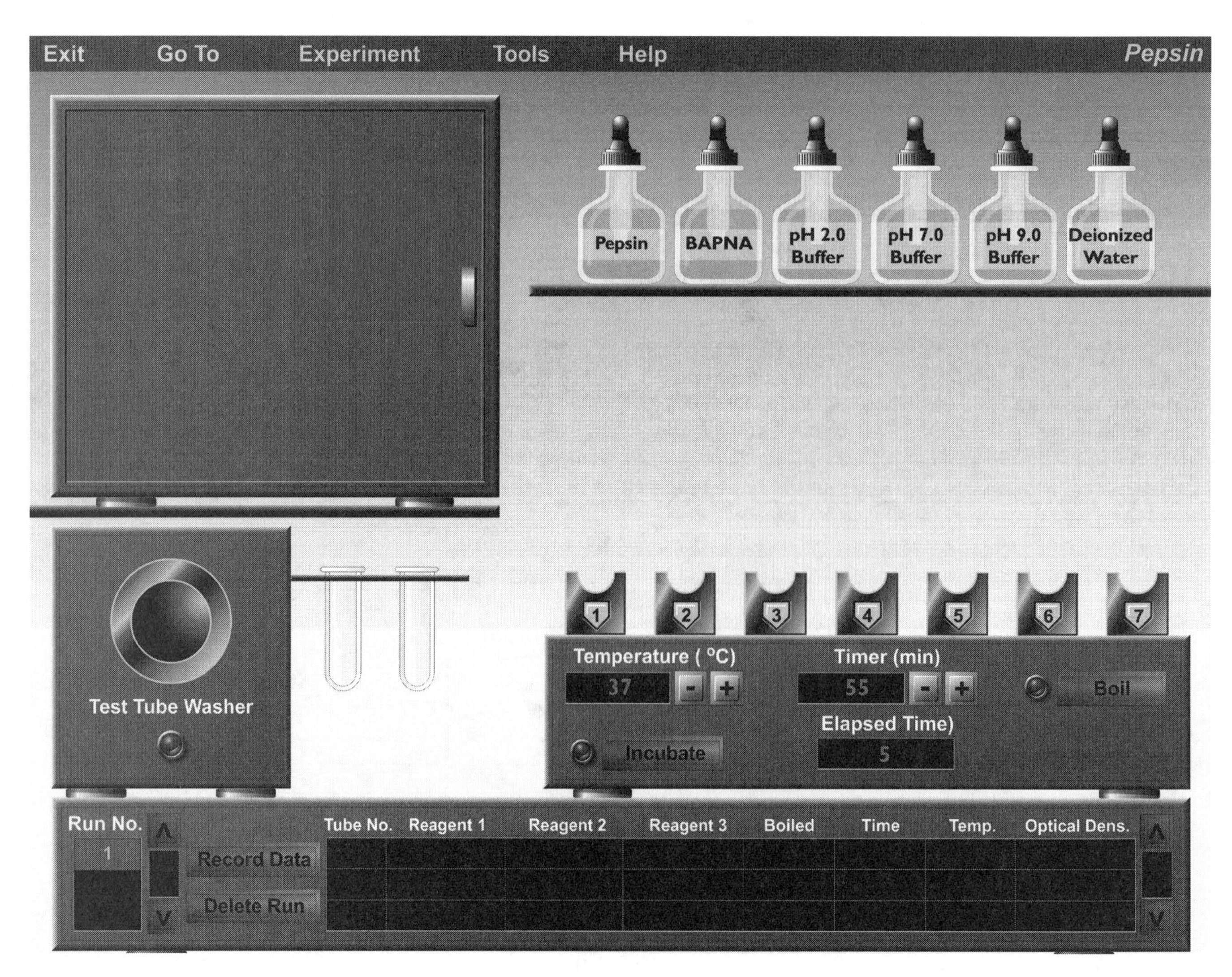

Figure 8.2 Opening screen of the Pepsin experiment.

ment, you will use BAPNA, a synthetic "protein" that is transparent and colorless when in solution. However, if an active, protein-digesting enzyme such as pepsin is present, the solution will become yellow. You can use this characteristic to detect pepsin activity: the solution turns yellow if the enzyme digests the BAPNA substrate; it remains colorless if pepsin is not active or not present. One advantage of using a synthetic substrate is that you do not need any additional indicator reagents to see enzyme activity.

Choose **Pepsin** from the **Experiment** menu. The opening screen will appear in a few seconds (Figure 8.2). The solutions shelf, test tube washer, and incubation equipment are the same as in the amylase experiment; only the solutions have changed.

Data displayed in the grid include the tube number, the three substances dispensed into each tube, a (+) or (−) mark indicating whether or not a sample was boiled, the time and temperature of the incubation, and the optical density measurement indicating enzyme assay results.

Activity 3: Assessing Protein Digestion by Pepsin

Pepsin Incubation

1. Individually drag six test tubes to the test tube holders in the incubation unit.

2. Prepare the tubes with the substances indicated in the chart below using the following method.

- Click and hold the mouse button on the dropper cap of the desired substance and drag the dropper cap to the top of the desired test tube.
- Release the mouse button to dispense the substance.

3. Once all tubes are prepared, click the number **(1)** under the first test tube. The tube will descend into the incubation unit. All other tubes should remain in the raised position.

4. Click **Boil** to boil tube 1. After boiling for a few moments, the tube will automatically rise.

5. Adjust the incubation temperature to 37°C and the timer to 60 min (compressed time) by clicking the (+) or (−) buttons.

6. Click **Incubate** to start the run. The incubation unit will gently agitate the test tube rack, evenly mixing the contents of all test tubes throughout the incubation. The computer is compressing the 60-minute time period into 60 seconds of real time. When the incubation time elapses, the test tube rack will automatically rise, and the doors to the assay cabinet will open.

Pepsin Assay

After the assay cabinet doors open, you will see an instrument called a spectrophotometer, which you will use to measure how much yellow dye was liberated by pepsin digestion of BAPNA. When a test tube is dragged to the holder in the spectrophotometer and the **Analyze** button is clicked, the instrument will shine a light through a specimen to measure the amount of light absorbed by the sample within the tube. The measure of the amount of light absorbed by the solution is known as its *optical density.* A colorless solution does not absorb light, whereas a colored solution has a relatively high light absorbance. For example, a colorless solution has an optical density of 0.0. A colored solution, however, absorbs some of the light emitted by the spectrophotometer, resulting in an optical density reading greater than zero.

In this experiment a yellow-colored solution is a direct indication of the amount of BAPNA digested by pepsin.

Pepsin Digestion of Protein

Tube no.	1	2	3	4	5	6
Additives	Pepsin BAPNA pH 2.0 buffer	Pepsin BAPNA pH 2.0 buffer	Pepsin Deionized water pH 2.0 buffer	Deionized water BAPNA pH 2.0 buffer	Pepsin BAPNA pH 7.0 buffer	Pepsin BAPNA pH 9.0 buffer
Incubation condition	Boil first, then incubate at 37°C	37°C	37°C	37°C	37°C	37°C
Optical density						

Although you can visually estimate the yellow color produced by pepsin digestion of BAPNA, the spectrophotometer precisely measures how much BAPNA digestion occurred in the experiment.

1. Click and hold the mouse on the first tube in the incubation unit and drag it to the holder in the spectrophotometer.
2. Release the mouse button to drop the tube into the holder.
3. Click **Analyze.** You will see light shining through the solution in the test tube as the spectrophotometer measures its optical density. The optical density of the sample will be displayed in the optical density window below the Analyze button.
4. Record the optical density in the chart.
5. Drag the tube to its original position in the incubation unit and release the mouse button.
6. Repeat steps 1 through 5 for the remaining test tubes in the incubation unit.
7. Click **Record Data** to display your results in the grid and retain your data in the computer's memory for later analysis. To repeat the experiment, you must drag all test tubes to the test tube washer and start again.
8. Answer the following questions, referring to the chart (or the data grid in the simulation) as necessary.

Which pH provided the highest pepsin activity? _______

Would pepsin be active in the mouth? Explain your answer.

____

____

How did the results of tube 1 compare with those of tube 2?

____

Tubes 1 and 2 contained the same substances. Explain why their optical density measurements were different.

____

____

If you had not run the tube 2 and 3 samples, what argument could be made against the statement "Pepsin digests BAPNA"?

____

____

What do you think would happen if you reduced the incubation time to 30 minutes? Use the simulation to help you answer this question if you are not sure.

____

____

What do you think would happen if you decreased the temperature of incubation to 10°C? Use the simulation to help you answer this question if you are not sure.

____

____

Click **Tools** → **Print Data** to print your recorded data. ■

Fat Digestion by Pancreatic Lipase and the Action of Bile

The treatment that fats and oils undergo during digestion in the small intestine is a bit more complicated than that of carbohydrates or proteins. Fats and oils require pretreatment with bile to physically emulsify the fats. As a result, two sets of reactions must occur.

First:

$$\text{Fats/oils} \xrightarrow[\text{(emulsification)}]{\text{bile}} \text{minute fat/oil droplets}$$

Then:

$$\text{Fat/oil droplets} \xrightarrow{\text{lipase}} \text{monoglycerides and fatty acids}$$

Lipase hydrolyzes fats and oils to their component monoglycerides and two fatty acids. Occasionally lipase hydrolyzes fats and oils to glycerol and three fatty acids.

The fact that some of the end products of fat digestion (fatty acids) are organic acids that decrease the pH provides an easy way to recognize that digestion is ongoing or completed. You will be using a pH meter in the assay cabinet to record the drop in pH as the test tube contents become acid.

Choose **Lipase** from the **Experiment** menu. The opening screen will appear in a few seconds (Figure 8.3). The solutions shelf, test tube washer, and incubation equipment are the same as in the previous two experiments; only the solutions have changed.

Data displayed in the grid include the tube number, the four reagents dispensed into each tube, a (+) or (−) mark indicating whether or not a sample was boiled, the time and temperature of the incubation, and the pH measurement indicating enzyme assay results.

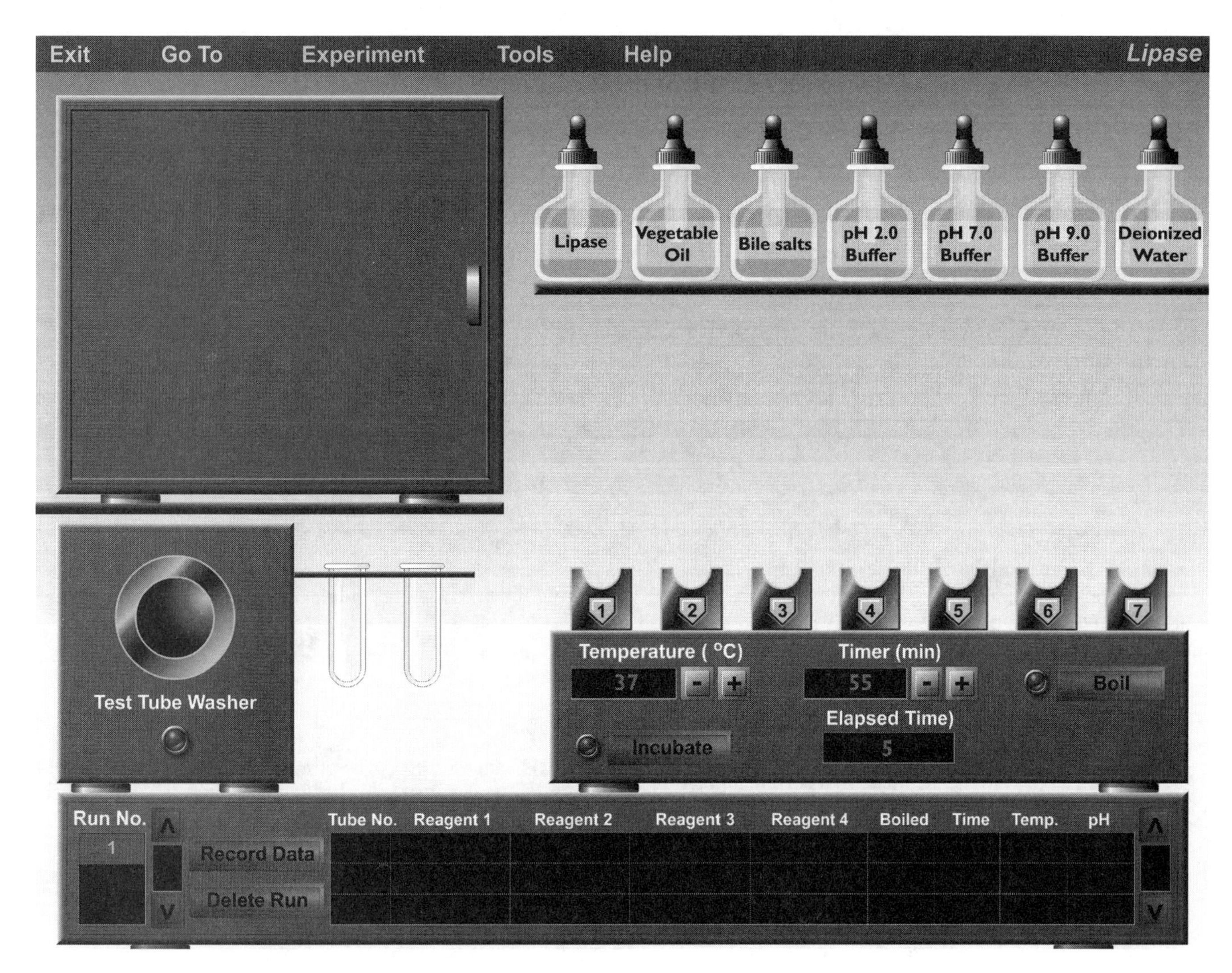

Figure 8.3 Opening screen of the Lipase experiment.

Activity 4:

Assessing Fat Digestion by Pancreatic Lipase and the Action of Bile

Lipase Incubation

1. Individually drag 6 test tubes to the test tube holders in the incubation unit.

2. Prepare the tubes with the solutions indicated in the chart on page 82 by using the following method.

- Click and hold the mouse button on the dropper cap of the desired substance.
- While holding the mouse button down, drag the dropper cap to the top of the desired test tube.
- Release the mouse button to dispense the substance.

3. Adjust the incubation temperature to 37°C and the timer to 60 min (compressed time) by clicking the (+) or (−) buttons.

4. Click **Incubate** to start the run. The incubation unit will gently agitate the test tube rack, evenly mixing the contents of all test tubes throughout the incubation. The computer is compressing the 60-minute time period into 60 seconds of real time. When the incubation time elapses, the test tube rack automatically rises, and the doors to the assay cabinet open.

Lipase Assay

After the assay cabinet doors open, you will see a pH meter that you will use to measure the relative acidity of your test solutions. When a test tube is dragged to the holder in the pH meter and the Measure pH button is clicked, a probe will descend into the sample, take a pH reading, and then retract. The pH of the sample will be displayed in the pH window below the Measure pH button. A solution containing fatty acids liberated from fat by the action of lipase will exhibit a lower pH than one without fatty acids.

Pancreatic Lipase Digestion of Fats and the Action of Bile

Tube no.	1	2	3	4	5	6
Additives	Lipase Vegetable oil Bile salts pH 7.0 buffer	Lipase Vegetable oil Deionized water pH 7.0 buffer	Lipase Deionized water Bile salts pH 9.0 buffer	Deionized water Vegetable oil Bile salts pH 7.0 buffer	Lipase Vegetable oil Bile salts pH 2.0 buffer	Lipase Vegetable oil Bile salts pH 9.0 buffer
Incubation condition	37°C	37°C	37°C	37°C	37°C	37°C
pH						

1. Click and hold the mouse on the first tube in the incubation unit and drag it to the holder in the pH meter. Release the mouse button to drop the tube into the holder.
2. Click **Measure pH.**
3. In the chart above, record the pH displayed in the pH window.
4. Drag the test tube in the pH meter to its original position in the incubation unit and release the mouse button.
5. Repeat steps 1 through 4 for the remaining test tubes in the incubation unit.
6. Click **Record Data** to display your results in the grid and retain your data in the computer's memory for later analysis. To repeat the experiment, you must drag all test tubes to the test tube washer and begin again.
7. Answer the following questions, referring to the chart (or the data grid in the simulation) as necessary.

Explain the difference in activity between tubes 1 and 2.

____

____

Can we determine if fat hydrolysis has occurred in tube 6?

_______ Explain your answer. ______________________

____

Which pH resulted in maximum lipase activity? _________

Is this method of assay sufficient to determine if the optimum activity of lipase is at pH 2.0? _______

In theory, would lipase be active in the mouth? _______

Would it be active in the stomach? _______

Explain your answers. ______________________

____

Based on the enzyme pH optima you determined, where in the body would we expect to find the enzymes in these experiments?

Amylase ______________________

Pepsin ______________________

Lipase ______________________

Click **Tools → Print Data** to print your recorded data. ■

Physical Processes: Mechanisms of Food Propulsion and Mixing

Although enzyme activity is an essential part of the overall digestion process, food must also be processed physically by churning and chewing, and moved by mechanical means along the tract if digestion and absorption are to be completed. Just about any time organs exhibit mobility, muscles are involved, and movements of and in the gastrointestinal tract are no exception. Although we tend to think only of smooth muscles for visceral activities, both skeletal and smooth muscles are necessary in digestion. This fact is demonstrated by the simple trials in the next activity. Obtain the following materials:

- Water pitcher
- Paper cups
- Stethoscope
- Alcohol swabs
- Disposable autoclave bag

Activity 5: Studying Mechanisms of Food Propulsion and Mixing

Deglutition (Swallowing)

Swallowing, or *deglutition,* which is largely the result of skeletal muscle activity, occurs in two phases: *buccal* (mouth) and *pharyngeal-esophageal.* The initial phase—the buccal—is voluntarily controlled and initiated by the tongue. Once begun, the process continues involuntarily in the pharynx and esophagus, through peristalsis, resulting in the delivery of the swallowed contents to the stomach.

1. While swallowing a mouthful of water, consciously note the movement of your tongue during the process. Record your observations.

____

____

____

2. Repeat the swallowing process while your laboratory partner watches the externally visible movements of your larynx. This movement is more obvious in a male, who has a larger Adam's apple. Record your observations.

____

____

What do these movements accomplish? ______________

____

____

3. Your lab partner should clean the ear pieces of a stethoscope with an alcohol swab and don the stethoscope. Then your lab partner should place the diaphragm of the stethoscope on your abdominal wall, approximately 1 inch below the xiphoid process and slightly to the left, to listen for sounds as you again take two or three swallows of water. There should be two audible sounds. The first sound occurs when the water splashes against the gastroesophageal sphincter. The second occurs when the peristaltic wave of the esophagus arrives at the sphincter and the sphincter opens, allowing water to gurgle into the stomach. Determine, as accurately as possible, the time interval between these two sounds and record it below.

Interval between arrival of water at the sphincter and the opening of the sphincter:

________ sec.

This interval gives a fair indication of the time it takes for the peristaltic wave to travel down the 10-inch-long esophagus. Actually the time interval is slightly less than it seems because pressure causes the sphincter to relax before the peristaltic wave reaches it.

Dispose of the used paper cup in the autoclave bag. ■

Segmentation and Peristalsis

Although several types of movement occur in the digestive tract organs, segmentation and peristalsis are most important as mixing and propulsive mechanisms.

Segmental movements are local constrictions of the organ wall that occur rhythmically. They serve mainly to mix the foodstuffs with digestive juices and to increase the rate of absorption by continually moving different portions of the chyme over adjacent regions of the intestinal wall. However, segmentation is an important means of food propulsion in the small intestine, and slow segmenting movements called haustral contractions are common in the large intestine.

Peristaltic movements are the major means of propelling food through most of the digestive viscera. Essentially they are waves of contraction followed by waves of relaxation that squeeze foodstuffs through the alimentary canal, and they are superimposed on segmental movements.

Histology Review Supplement

Turn to p. P-148 for a review of digestive tissue.

Exercise

9 Renal Physiology—The Function of the Nephron: Computer Simulation

Objectives

1. To define:
   glomerulus, glomerular capsule, renal corpuscle, renal tubule, nephron, proximal convoluted tubule, loop of Henle, and *distal convoluted tubule*
2. To describe the blood supply to each nephron.
3. To identify the regions of the nephron involved in glomerular filtration and tubular reabsorption.
4. To study the factors affecting glomerular filtration.
5. To explore the concept of carrier transport maximum.
6. To understand how the hormones aldosterone and ADH affect the function of the kidney.
7. To describe how the kidneys can produce urine that is four times more concentrated than the blood.

Metabolism produces wastes that must be eliminated from the body. This excretory function is the job of the renal system, most importantly the paired kidneys. Each kidney consists of about one million nephrons that carry out two crucial services, blood filtration and fluid processing.

Microscopic Structure and Function of the Kidney

Each of the million or so **nephrons** in each kidney is a microscopic tubule consisting of two major parts: a glomerulus and a renal tubule. The **glomerulus** is a tangled capillary knot that filters fluid from the blood into the lumen of the renal tubule. The function of the **renal tubule** is to process that fluid, also called the **filtrate.** The beginning of the renal tubule is an enlarged end called the **glomerular capsule,** which surrounds the glomerulus and serves to funnel the filtrate into the rest of the renal tubule. Collectively, the glomerulus and the glomerular capsule are called the **renal corpuscle.**

As the rest of the renal tubule extends from the glomerular capsule, it becomes twisted and convoluted, then dips sharply down to form a hairpin loop, and then coils again before entering a collecting duct. Starting at the glomerular capsule, the anatomical parts of the renal tubule are as follows: the **proximal convoluted tubule,** the **loop of Henle** (nephron loop), and the **distal convoluted tubule.**

Two arterioles supply each glomerulus: an afferent arteriole feeds the glomerular capillary bed and an efferent arteriole drains it. These arterioles are responsible for blood flow through the glomerulus. Constricting the afferent arteriole lowers the downstream pressure in the glomerulus, whereas constricting the efferent arteriole will increase the pressure in the glomerulus. In addition, the diameter of the efferent arteriole is smaller than the diameter of the afferent arteriole, restricting blood flow out of the glomerulus. Consequently, the pressure in the glomerulus forces fluid through the endothelium of the glomerulus into the lumen of the surrounding glomerular capsule. In essence, everything in the blood except the cells and proteins are filtered through the glomerular wall. From the capsule, the filtrate moves into the rest of the renal tubule for processing. The job of the tubule is to reabsorb all the beneficial substances from its lumen while allowing the wastes to travel down the tubule for elimination from the body.

The nephron performs three important functions to process the filtrate into urine: glomerular filtration, tubular reabsorption, and tubular secretion. **Glomerular filtration** is a passive process in which fluid passes from the lumen of the glomerular capillary into the glomerular capsule of the renal tubule. **Tubular reabsorption** moves most of the filtrate back into the blood, leaving principally salt water plus the wastes in the lumen of the tubule. Some of the desirable or needed solutes are actively reabsorbed, and others move passively from the lumen of the tubule into the interstitial spaces. **Tubular secretion** is essentially the reverse of tubular reabsorption and is a process by which the kidneys can rid the blood of additional unwanted substances such as creatinine and ammonia.

The reabsorbed solutes and water that move into the interstitial space between the nephrons need to be returned to the blood, or the kidneys will rapidly swell like balloons. The peritubular capillaries surrounding the renal tubule reclaim the reabsorbed substances and return them to general circulation. Peritubular capillaries arise from the efferent arteriole exiting the glomerulus and empty into the veins leaving the kidney.

Simulating Glomerular Filtration

This computerized simulation allows you to explore one function of a single simulated nephron, glomerular filtration. The concepts you will learn by studying a single nephron can then be applied to understand the function of the kidney as a whole.

Choose **Renal System Physiology** from the main menu. The opening screen for the Simulating Glomerular Filtration experiment will appear in a few seconds (Figure 9.1). The main features on the screen when the program starts are a simulated blood supply at the left side of the screen, a simulated nephron within a supporting tank on the right side, and a data control unit at the bottom of the display.

The left beaker is the "blood" source representing the general circulation supplying the nephron. The "blood pressure" in the beaker is adjustable by clicking the (+) and (–)

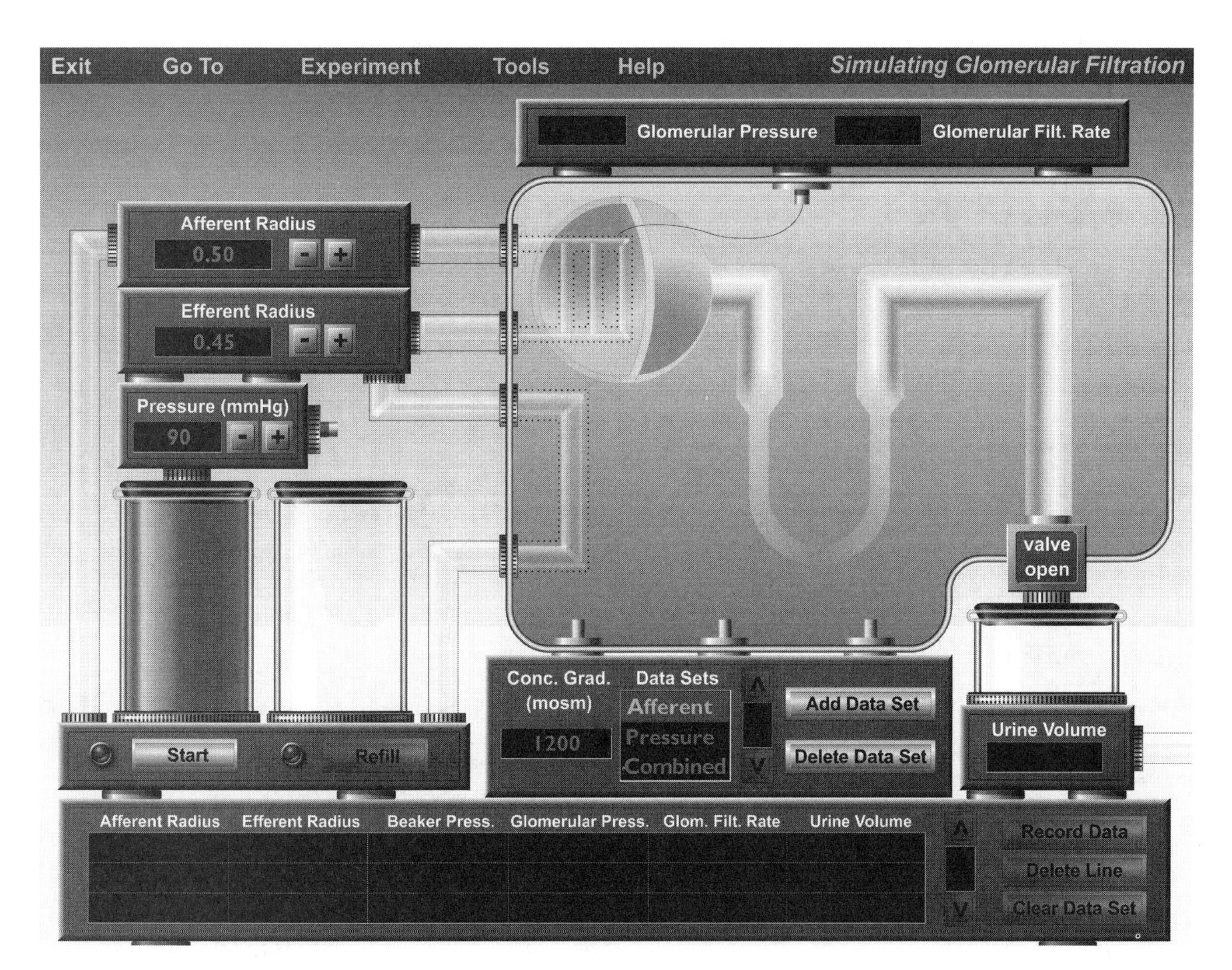

Figure 9.1 Opening screen of the Simulating Glomerular Filtration experiment.

buttons on top of the beaker. A tube with an adjustable radius called the *afferent flow* tube connects the left beaker to the simulated glomerulus. Another adjustable tube called the *efferent flow* tube drains the glomerulus. The afferent flow tube represents the afferent arteriole feeding the glomerulus of each nephron, and the efferent flow tube represents the efferent arteriole draining the glomerulus. The outflow of the nephron empties into a collecting duct, which in turn drains into another small beaker at the bottom right part of the screen. Clicking the valve at the end of the collecting duct stops the flow of fluid through the nephron and collecting duct.

The Glomerular Pressure window on top of the nephron tank displays the pressure within the glomerulus. The Glomerular Filt. Rate window indicates the flow rate of the fluid moving from the lumen of the glomerulus into the lumen of the renal tubule.

The concentration gradient bathing the nephron is fixed at 1200 mosm. Clicking **Start** begins the experiment. Clicking **Refill** resets the equipment to begin another run.

The equipment in the lower part of the screen is called the *data control unit.* This equipment records and displays data you accumulate during the experiments. The data set for the first experiment (Afferent) is highlighted in the **Data Sets** window. You can add or delete a data set by clicking the appropriate button to the right of the Data Sets window. When you click **Record Data,** your data is recorded in the computer's memory and is displayed in the data grid. Data displayed in the data grid include the Afferent Radius, Efferent Radius, Beaker Pressure, Glomerular Pressure, Glomerular Filtration Rate, and the Urine Volume. Clicking **Delete Line** allows you to discard data values for a single run, and clicking **Clear Data Set** erases the entire experiment to allow you to start over.

If you need help identifying any piece of equipment, choose **Balloons On** from the Help menu and move the mouse pointer onto any piece of equipment visible on the computer's screen. As the pointer touches the object, a pop-up window appears identifying the equipment. To close the pop-up window, move the mouse pointer away from the equipment. Choose **Balloons Off** to turn off this help feature.

Activity 1: Investigating the Effect of Flow Tube Radius on Glomerular Filtration

Your first experiment will examine the effects of flow tube radii and pressures on the rate of glomerular filtration. Click **Start** to see the on-screen action. Continue when you understand how the simulation operates. Click **Refill** to reset the experiment.

1. The **Afferent** line in the Data Sets window of the data control unit should be highlighted in bright blue. If it is not, choose it by clicking the **Afferent** line. The data control unit will now record filtration rate variations due to changing afferent flow tube radius.

2. If the data grid is not empty, click **Clear Data Set** to discard all previous data.

3. Adjust the afferent radius to 0.35 mm, and the efferent radius to 0.40 mm by clicking the appropriate buttons.

4. If the left beaker is not full, click **Refill.**

5. Keep the beaker pressure at 90 mm Hg during this part of the experiment.

6. Click **Start** and watch the blood flow. Simultaneously, filtered fluid will be moving through the nephron and into the collecting duct. The Glomerular Filtration Rate window will display the fluid flow rate into the renal tubule when the left beaker has finished draining.

7. Now click **Record Data** to record the current experiment data in the data grid. Click **Refill** to replenish the left beaker and prepare the nephron for the next run.

8. Increase the afferent radius in 0.05-mm increments and repeat steps 6 through 8 until the maximum radius (0.60 mm) is achieved. Be sure to click **Record Data** after each trial. If you make an error and want to delete a single value, click the data line in the data grid and then click **Delete Line.**

What happens to the glomerular filtration rate as the afferent radius is increased?

____

____

Predict the effect of increasing or decreasing the efferent radius on glomerular filtration rate. Use the simulation to reach an answer if you are not sure.

____

______________________________ ■

Activity 2: Studying the Effect of Pressure on Glomerular Filtration

Both the blood pressure supplying the glomerulus and the pressure in the renal tubule have a significant impact on the glomerular filtration rate. In this activity, the data control unit will record filtration rate variations due to changing pressure.

1. Click the **Pressure** line in the Data Sets window of the data control unit.

2. If the data grid is not empty, click **Clear Data Set** to discard all previous data.

3. If the left beaker is not full, click **Refill.**

4. Adjust the pressure in the left beaker to 70 mm Hg by clicking the appropriate button.

5. During this part of the experiment, maintain the afferent flow tube radius at 0.55 mm and the efferent flow tube radius at 0.45 mm.

6. Click **Start** and watch the blood flow. Filtrate will move through the nephron into the collecting duct. At the end of the

run, the Glomerular Filtration Rate window will display the filtrate flow rate into the renal tubule.

7. Now click **Record Data** to record the current experiment data in the data grid. Click **Refill** to replenish the left beaker.

8. Increase the pressure in the upper beaker in increments of 10 mm Hg and repeat steps 6 through 8 until the maximum pressure (100 mm Hg) is achieved. Be sure to click **Record Data** after each trial. If you make an error and want to delete a single value, click the data line in the data grid and then click **Delete Line.**

What happened to the glomerular filtration rate as the beaker pressure was increased?

____

Explain your answer.

____

____

____

____ ■

Activity 3:
Assessing Combined Effects on Glomerular Filtration

So far, you have examined the effects of flow tube radius and pressure on glomerular filtration rate. In this experiment you will be altering both variables to explore the combined effects on glomerular filtration rate and to see how one can compensate for the other to maintain an adequate glomerular filtration rate.

1. Click **Combined** in the Data Sets window of the data control unit.

2. If the data grid is not empty, click **Clear Data Set** to discard all previous data.

3. If the left beaker is not full, click **Refill.**

4. Set the starting conditions at

- 100 mm Hg beaker pressure
- 0.55 mm afferent radius
- 0.45 mm efferent radius

5. Click **Start.**

6. Now click **Record Data** to record the current baseline data in the data grid.

7. Click **Refill.**

You will use this baseline data to compare a run in which the valve at the end of the collecting duct is in the open position with a run in which the valve is in the closed position.

8. Use the simulation and your knowledge of basic renal anatomy to arrive at answers to the following questions. Be sure to click **Record Data** after each trial. If you make an error and want to delete a single value, click the data line in the data grid and then click **Delete Line.**

Click **valve open** on the end of the collecting duct. Note that it now reads **valve closed.** Click **Start** and allow the run to complete. How does this run compare to the runs in which the valve was open?

____

Expanding on this concept, what might happen to total glomerular filtration and therefore urine production in a human kidney if all of its collecting ducts were totally blocked?

____

____

Would kidney function as a whole be affected if a single nephron was blocked? Explain.

____

____

Would the kidney be functioning if glomerular filtration was zero? Explain.

____

____

Explain how the body could increase glomerular filtration rate in a human kidney.

____

____

____

____

If you increased the pressure in the beaker, what other condition(s) could you adjust to keep the glomerular filtration rate constant?

____

____

Click **Tools** → **Print Data** to print your recorded data. ■

Simulating Urine Formation

This part of the computer simulation allows you to explore some aspects of urine formation by manipulating the interstitial solute concentration. Other activities include investigating the effects of aldosterone and ADH (antidiuretic hormone), and the role that glucose carrier proteins play in renal function.

Choose **Simulating Urine Formation** from the **Experiment** menu. The opening screen will appear in a few seconds (Figure 9.2). The basic features on the screen when the program starts are similar to the glomerular filtration screen. Most of the vascular controls have been moved off-screen to the left because they will not be needed in this set of experiments. Additional equipment includes a *supplies shelf* at the right side of the screen, a *glucose carrier control* located at the top of the nephron tank, and a concentration probe at the bottom left part of the screen.

The maximum concentration of the "interstitial gradient" to be dispensed into the tank surrounding the nephron is adjusted by clicking the (+) and (–) buttons next to the Conc. Grad. window. Click **Dispense** to fill the tank through the jets at the bottom of the tank with the chosen solute gradient. Click **Start** to begin a run. After a run completes, the concentration probe can be clicked and dragged over the nephron to display the solute concentration within.

Hormone is dispensed by dragging a hormone bottle cap to the gray cap button in the nephron tank at the top of the collecting duct and then letting go of the mouse button.

The (+) and (–) buttons in the glucose carrier control are used to adjust the number of glucose carriers that will be inserted into the simulated proximal convoluted tubule when the Add Carriers button is clicked.

Data displayed in the data grid will depend on which experiment is being conducted. Clicking **Delete Line** allows you to discard data values for a single run, and clicking **Clear Data Set** erases the entire experiment to allow you to start over.

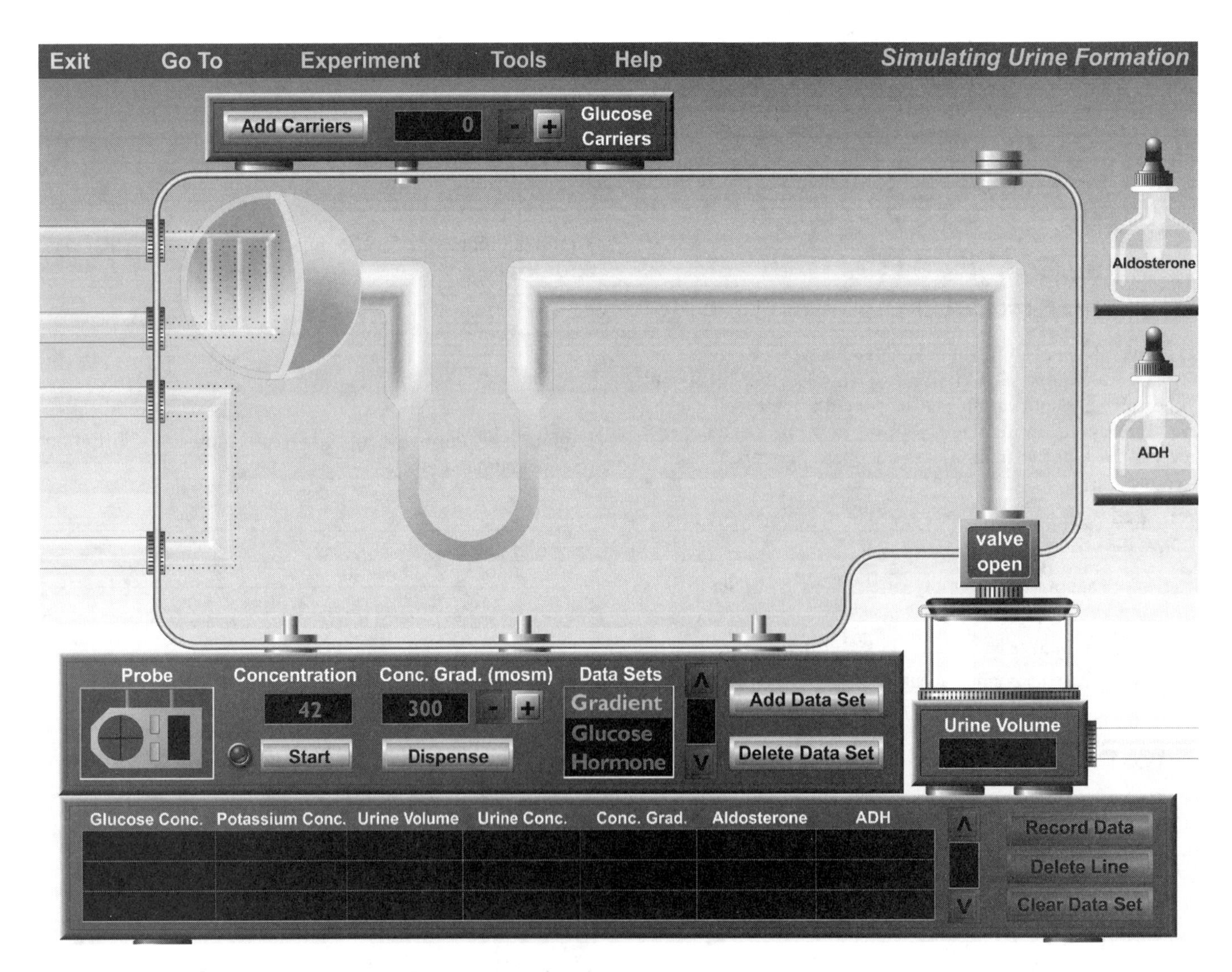

Figure 9.2 Opening screen of the Simulating Urine Formation experiment.

Activity 4:
Exploring the Role of the Solute Gradient on Maximum Urine Concentration Achievable

In the process of urine formation, solutes and water move from the lumen of the nephron into the interstitial spaces. The passive movement of solutes and water from the lumen of the renal tubule into the interstitial spaces relies in part on the total solute gradient surrounding the nephron. When the nephron is permeable to solutes or water, an equilibrium will be reached between the interstitial fluid and the contents of the nephron. Antidiuretic hormone (ADH) increases the water permeability of the distal convoluted tubule and the collecting duct, allowing water to flow to areas of higher solute concentration, usually from the lumen of the nephron into the surrounding interstitial area. You will explore the process of passive reabsorption in this experiment. While doing this part of the simulation, assume that when ADH is present the conditions favor the formation of the most concentrated urine possible.

1. **Gradient** in the Data Sets window of the data control unit should be highlighted in bright blue. If it is not then click **Gradient.**

2. If the data grid is not empty, click **Clear Data Set** to discard all previous data.

3. Click and hold the mouse button on the ADH bottle cap and drag it to the gray cap at the top right side of the nephron tank. Release the mouse button to dispense ADH onto the collecting duct.

4. Adjust the maximum total solute concentration of the gradient **(Conc. Grad.)** to 300 mosm by clicking the appropriate button. Because the blood solute concentration is also 300 mosm, there is no osmotic difference between the lumen of the nephron and the surrounding interstitial fluid.

5. Click **Dispense.**

6. Click **Start** to begin the experiment. Filtrate will move through the nephron and then drain into the beaker below the collecting duct.

7. While the experiment is running, watch the Probe. When it turns red, click and hold the mouse on it and drag it to the urine beaker. Observe the total solute concentration in the Concentration window.

8. Now click **Record Data** to record the current experiment data in the data grid.

9. Increase the maximum concentration of the gradient in 300-mosm increments and repeat steps 3 through 8 until 1200 mosm is achieved. Be sure to click **Record Data** after each trial. If you make an error and want to delete a single value, click the data line in the data grid and then click **Delete Line.**

What happened to the urine concentration as the gradient concentration was increased?

____

____

What factor limits the maximum possible urine concentration?

____

____

The solute concentration of the blood is about 300 mosm, and the highest interstitial solute concentration in a human kidney is about 1200 mosm. This means that the maximum urine solute concentration is about four times that of the blood. What would be the maximum possible urine concentration if the maximum interstitial solute concentration were 3000 mosm instead of 1200 mosm? Explain. (Use the simulation to arrive at an answer if you are not sure.)

____

____

____

____ ■

Activity 5:
Studying the Effect of Glucose Carrier Proteins on Glucose Reabsorption

Because carrier proteins are needed to move glucose from the lumen of the nephron into the interstitial spaces, there is a limit to the amount of glucose that can be reabsorbed. When all glucose carriers are bound with the glucose they are transporting, excess glucose is eliminated in urine. In this experiment, you will examine the effect of varying the number of glucose transport proteins in the proximal convoluted tubule.

1. Click **Glucose** in the Data Sets window of the data control unit.

2. If the data grid is not empty, click **Clear Data Set** to discard all previous data.

3. Set the concentration gradient **(Conc. Grad.)** to 1200 mosm.

4. Click **Dispense.**

5. Adjust the number of glucose carriers to 100 (an arbitrary figure) by clicking the appropriate button.

6. Click **Add Carriers.** This action inserts the specified number of glucose carrier proteins per unit area into the membrane of the proximal convoluted tubule.

7. Click **Start** to begin the run after the carriers have been added.

8. Click **Record Data** to record the current experiment data in the data grid. Glucose presence in the urine will be displayed in the data grid.

9. Now increase the number of glucose carrier proteins in the proximal convoluted tubule in increments of 100 glucose carriers and repeat steps 6 through 8 until the maximum number of glucose carrier proteins (500) is achieved. Be sure

to click **Record Data** after each trial. If you make an error and want to delete a single value, click the data line in the data grid and then click **Delete Line.**

What happened to the amount of glucose present in the urine as the number of glucose carriers was increased?

____

The amount of glucose present in normal urine is minimal because there are normally enough glucose carriers present to handle the "traffic." Predict the consequence in the urine if there was more glucose than could be transported by the available number of glucose carrier proteins.

____

____

Explain why we would expect to find glucose in the urine of a diabetic person.

____

_______________ ■

Activity 6:
Testing the Effect of Hormones on Urine Formation

The concentration of the urine excreted by our kidneys changes depending on our immediate needs. For example, if a person consumes a large quantity of water, the excess water will be eliminated, producing dilute urine. On the other hand, under conditions of dehydration, there is a clear benefit in being able to produce urine as concentrated as possible, thereby retaining precious water. Although the medullary gradient makes it possible to excrete concentrated urine, urine dilution or concentration is ultimately under hormonal control. In this experiment, you will investigate the effects of two different hormones on renal function, aldosterone produced by the adrenal gland and ADH manufactured by the hypothalamus and stored in the posterior pituitary gland. Aldosterone works to reabsorb Na^+ (and thereby water) at the expense of losing K^+. Its site of action is the distal convoluted tubule. ADH makes the distal tubule and collecting duct more permeable to water, thereby allowing the body to reabsorb more water from the filtrate when it is present.

1. Click **Hormone** in the Data Sets window of the data control unit.

2. If the data grid is not empty, click **Clear Data Set** to discard all previous data.

3. During this part of the experiment, keep the concentration gradient at 1200 mosm.

4. Click **Dispense** to add the gradient and then click **Start** to begin the experiment.

5. Now click **Record Data** to record the current experiment data in the data grid.

You will use this baseline data to compare with the conditions of the filtrate under the control of the two hormones.

6. Keeping all experiment conditions the same as before, do the following:

- Drag the aldosterone bottle cap to the gray cap on the top right side of the nephron tank and release the mouse to automatically dispense aldosterone into the tank surrounding the distal convoluted tubule and collecting duct.
- Click **Start** and allow the run to complete.
- Click **Record Data.**

In this run, how does the volume of urine differ from the previously measured baseline volume?

____

____

Explain the difference in the total amount of potassium in the urine between this run and the baseline run.

____

____

7. Drag the ADH bottle cap to the gray cap on the top right side of the nephron tank and release it to dispense ADH.

- Click **Start** and allow the run to complete
- Click **Record Data.**

In this run, how does the volume of urine differ from the baseline measurement?

____

____

Is there a difference in the total amount of potassium in this run and the total amount of potassium in the baseline run? Explain. (Hint: the urine volume with ADH present is about one-tenth the urine volume when it is not present.)

____

____

Are the effects of aldosterone and ADH similar or antagonistic? _______________

Consider this situation: we want to reabsorb sodium ions but do not want to increase the volume of the blood by reabsorbing water from the filtrate. Assuming that aldosterone and ADH are both present, how would you adjust the hormones to accomplish the task?

____

____

____

____

If the interstitial gradient ranged from 300 mosm to 3000 mosm and ADH was not present, what would be the maximum possible urine concentration? ______________

Click **Tools** → **Print Data** to print your recorded data. ■

Histology Review Supplement

Turn to p. P-150 for a review of renal tissue.

Exercise

10 Acid-Base Balance: Computer Simulation

Objectives

1. To define *pH* and identify the normal range of human blood pH levels.
2. To define *acid* and *base*, and explain what characterizes each of the following: *strong acid, weak acid, strong base, weak base.*
3. To explain how chemical and physiological buffering systems help regulate the body's pH levels.
4. To define the conditions of *acidosis* and *alkalosis.*
5. To explain the difference between *respiratory acidosis and alkalosis* and *metabolic acidosis and alkalosis.*
6. To understand the causes of respiratory acidosis and alkalosis.
7. To explain how the renal system compensates for respiratory acidosis and alkalosis.
8. To understand the causes of metabolic acidosis and alkalosis.
9. To explain how the respiratory system compensates for metabolic acidosis and alkalosis.

The term **pH** is used to denote the hydrogen ion concentration $[H^+]$ in body fluids. pH values are the reciprocal of $[H^+]$ and follow the formula

$$pH = \log(1/[H^+])$$

At a pH of 7.4, $[H^+]$ is about 40 nanomolars (n*M*) per liter. Because the relationship is reciprocal, $[H^+]$ is higher at *lower* pH values (indicating higher acid levels) and lower at *higher* pH values (indicating lower acid levels).

The pH of a body's fluids is also referred to as its **acid-base balance.** An **acid** is a substance that releases H^+ in solution (such as in body fluids). A **base,** often a hydroxyl ion (OH^-) or bicarbonate ion (HCO_3^-), is a substance that binds to H^+. A *strong acid* is one that completely dissociates in solution, releasing all of its hydrogen ions and thus lowering the solution's pH level. A *weak acid* dissociates incompletely and does not release all of its hydrogen ions in solution. A *strong base* has a strong tendency to bind to H^+, which has the effect of raising the pH value of the solution. A *weak base* binds less of the H^+, having a lesser effect on solution pH.

The body's pH levels are very tightly regulated. Blood and tissue fluids normally have pH values between 7.35 and 7.45. Under pathological conditions, blood pH values as low as 6.9 or as high as 7.8 have been recorded; however, values higher or lower than these cannot sustain human life. The narrow range of 7.35–7.45 is remarkable when one considers the vast number of biochemical reactions that take place in the body. The human body normally produces a large amount of H^+ as the result of metabolic processes, ingested acids, and the products of fat, sugar, and amino acid metabolism. The regulation of a relatively constant internal pH environment is one of the major physiological functions of the body's organ systems.

To maintain pH homeostasis, the body utilizes both *chemical* and *physiological* buffering sytems. Chemical buffers are composed of a mixture of weak acids and weak bases. They help regulate body pH levels by binding H^+ and removing it from solution as its concentration begins to rise, or releasing H^+ into solution as its concentration begins to fall. The body's three major chemical buffering systems are the *bicarbonate, phosphate,* and *protein buffer systems.* We will not focus on chemical buffering systems in this lab, but keep in mind that chemical buffers are the fastest form of compensation and can return pH to normal levels within a fraction of a second.

The body's two major physiological buffering systems are the renal and respiratory systems. The renal system is the slower of the two, taking hours to days to do its work. The respiratory system usually works within minutes, but cannot handle the amount of pH change that the renal system can. These physiological buffer systems help regulate body pH by controlling the output of acids, bases, or CO_2 from the body. For example, if there is too much acid in the body, the renal system may respond by excreting more H^+ from the body in urine. Similarly, if there is too much carbon dioxide in the blood, the respiratory system may respond by breathing faster to expel the excess carbon dioxide. Carbon dioxide levels have a direct effect on pH levels because the addition of carbon dioxide to the blood results in the generation of more H^+. The following reaction shows what happens in the respiratory system when carbon dioxide combines with water in the blood:

$$H_2O + CO_2 \rightleftarrows \underset{\text{carbonic acid}}{H_2CO_3} \rightleftarrows H^+ + \underset{\text{bicarbonate ion}}{HCO_3^-}$$

This is a reversible reaction and is useful for remembering the relationships between CO_2 and H^+. Note that as more CO_2 accumulates in the blood (which frequently is caused by reduced gas exchange in the lungs), the reaction moves to the right and more H^+ is produced, lowering the pH:

$$H_2O + \mathbf{CO_2} \rightarrow \underset{\text{carbonic acid}}{H_2CO_3} \rightarrow \mathbf{H^+} + \underset{\text{bicarbonate ion}}{HCO_3^-}$$

Conversely, as $[H^+]$ increases, more carbon dioxide will be present in the blood:

$$H_2O + \mathbf{CO_2} \leftarrow \underset{\text{carbonic acid}}{H_2CO_3} \leftarrow \mathbf{H^+} + \underset{\text{bicarbonate ion}}{HCO_3^-}$$

Disruptions of acid-base balance occur when the body's pH levels fall below or above the normal pH range of 7.35–7.45. When pH levels fall below 7.35, the body is said to be in a state of **acidosis.** When pH levels rise above 7.45, the body is said to be in a state of **alkalosis. Respiratory acidosis** and **respiratory alkalosis** are the result of the respiratory system accumulating too much or too little carbon dioxide in the blood. **Metabolic acidosis** and **metabolic alkalosis** refer to all other conditions of acidosis and alkalosis (i.e., those not caused by the respiratory system). The experiments in this lab will focus on these disruptions of acid-base balance, and on the physiological buffer systems (renal and respiratory) that compensate for such imbalances.

Respiratory Acidosis and Alkalosis

Respiratory acidosis is the result of impaired respiration, or *hypoventilation,* which leads to the accumulation of too much carbon dioxide in the blood. The causes of impaired respiration include airway obstruction, depression of the respiratory center in the brain stem, lung disease, and drug overdose. Recall that carbon dioxide acts as an acid by forming carbonic acid when it combines with water in the body's blood. The carbonic acid then forms hydrogen ions plus bicarbonate ions:

$$H_2O + \mathbf{CO_2} \rightarrow \underset{\text{carbonic acid}}{H_2CO_3} \rightarrow \mathbf{H^+} + \underset{\text{bicarbonate ion}}{HCO_3^-}$$

Because hypoventilation results in elevated carbon dioxide levels in the blood, the H^+ levels increase, and the pH value of the blood decreases.

Respiratory alkalosis is the condition of too little carbon dioxide in the blood. It is commonly the result of traveling to a high altitude (where the air contains less oxygen) or hyperventilation, which may be brought on by fever or anxiety. Hyperventilation removes more carbon dioxide from the blood, reducing the amount of H^+ in the blood and thus increasing the blood's pH level.

In this first set of activities, we focus on the causes of respiratory acidosis and alkalosis. Follow the instructions in the Getting Started section on pp. P-2 and P-3 in the PhysioEx Introduction to start PhysioEx. From the Main Menu, select **Acid-Base Balance.** You will see the opening screen for "Respiratory Acidosis/Alkalosis" (Figure 10.1). If you have already completed PhysioEx Exercise 7 on respiratory system mechanics, this screen should look familiar. At the left is a pair of simulated lungs, which look like balloons, connected by a tube that looks like an upside-down Y. Air flows in and out of this tube, which simulates the trachea and other air passageways into the lungs. Beneath the "lungs" is a black platform simulating the diaphragm. The long, U-shaped tube containing red fluid represents blood flowing through the lungs. At the top left of the U-shaped tube is a pH meter that will measure the pH level of the blood once the experiment is begun (experiments are begun by clicking the **Start** button at the left of the screen). To the right is an oscilloscope monitor, which will graphically display respiratory volumes. Note that respiratory volumes are measured in liters (l) along the Y-axis, and time in seconds is measured along the X-axis. Below the monitor are three buttons: **Normal Breathing, Hyperventilation,** and **Rebreathing.** Clicking any one of these buttons will induce the given pattern of breathing. Next to these buttons are three data displays for P_{CO_2} (partial pressure of carbon dioxide)—these will give us the levels of carbon dioxide in the blood over the course of an experimental run. At the very bottom of the screen is the data collection grid, where you may record and view your data after each activity.

Activity 1: Normal Breathing

To get familiarized with the equipment, as well as to obtain baseline data for this experiment, we will first observe what happens during normal breathing.

1. Click **Start.** Notice that the **Normal Breathing** button dims, indicating that the simulated lungs are "breathing" normally. Also notice the reading in the pH meter at the top left, the readings in the P_{CO_2} displays, and the shape of the trace that starts running across the oscilloscope screen. As the trace runs, record the readings for pH at each of the following times:

At 20 seconds, pH = ______________

At 40 seconds, pH = ______________

At 60 seconds, pH = ______________

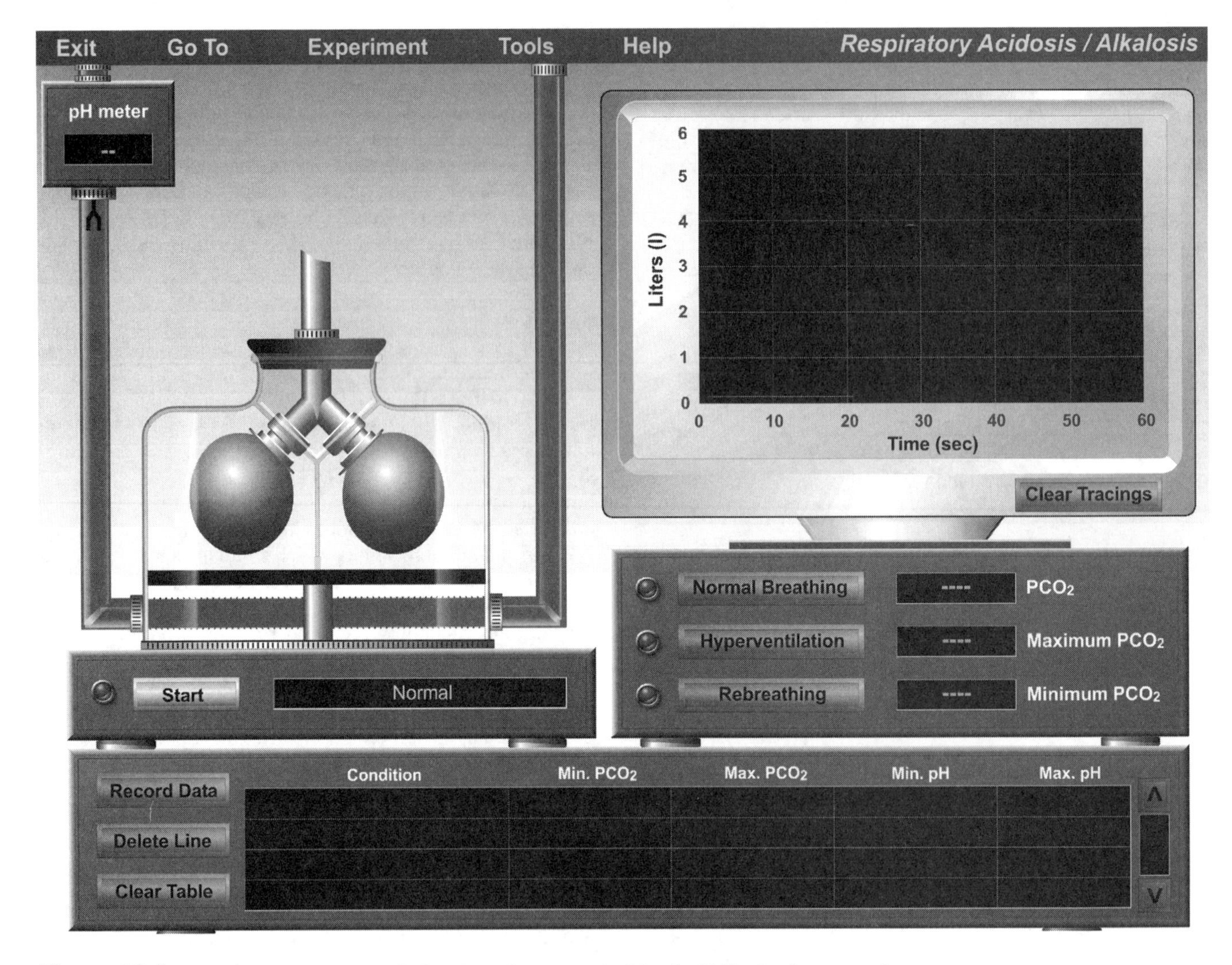

Figure 10.1 Opening screen of the Respiratory Acidosis/Alkalosis experiment.

2. Allow the trace to run all the way to the right side of the oscilloscope screen. At this point, the run will automatically end.

3. Click **Record Data** at the bottom left to record your results.

4. If you have printer access, click **Tools** at the top of the screen and select **Print Graph.** Otherwise, manually sketch what you see on the oscilloscope screen.

5. Click **Clear Tracings** to clear the oscilloscope screen.

Did the pH level of the blood change at all during normal breathing? If so, how?

____

Was the pH level always within the "normal" range for the human body?

____

Did the P_{CO_2} level change during the course of normal breathing? If so, how?

____ ■

Activity 2a: Hyperventilation—Run 1

Next, we will observe what happens to pH and carbon dioxide levels in the blood during hyperventilation.

1. Click **Start.** Allow the normal breathing trace to run for 10 seconds; then at the 10-second mark, click **Hyperventilation.** Watch the pH meter display, as well as the readings in the P_{CO_2} displays and the shape of the trace. As the trace runs, record the readings for pH at each of the following times:

At 20 seconds, pH = ______________

At 40 seconds, pH = ______________

At 60 seconds, pH = ______________

2. Allow the trace to run all the way across the oscilloscope screen and end.
3. Click **Record Data.**
4. If you have printer access, click **Tools** at the top of the screen and select **Print Graph.** Otherwise, manually sketch what you see on the oscilloscope screen on a separate sheet of paper.
5. Click **Clear Tracings** to clear the oscilloscope screen.

Did the pH level of the blood change at all during this run? If so, how?

____

Was the pH level always within the "normal" range for the

human body? ______________________________

If not, when was the pH value outside of the normal range, and what acid-base imbalance did this pH value indicate?

____

Did the P_{CO_2} level change during the course of this run? If so, how?

____

If you observed an acid-base imbalance during this run, how would you expect the renal system to compensate for this condition?

____

How did the hyperventilation trace differ from the trace for normal breathing? Did the tidal volumes change?

____

____

What might cause a person to hyperventilate?

____ ■

Activity 2b: Hyperventilation—Run 2

This activity is a variation on Activity 2a.

1. Click **Start.** Allow the normal breathing trace to run for 10 seconds, then click **Hyperventilation** at the 10-second mark. Allow the hyperventilation trace to run for 10 seconds, then click **Normal Breathing** at the 20-second mark. Allow the trace to finish its run across the oscilloscope screen. Observe the changes in the pH meter and the P_{CO_2} displays.
2. Click **Record Data.**
3. If you have printer access, click **Tools** at the top of the screen and select **Print Graph.** Otherwise, manually sketch what you see on the oscilloscope screen.
4. Click **Clear Tracings** to clear the oscilloscope screen.

What happened to the trace after the 20-second mark when you stopped the hyperventilation? Did the breathing return to normal immediately? Explain your observation.

____

____

____

____ ■

Activity 3: Rebreathing

Rebreathing is the action of breathing in air that was just expelled from the lungs. Breathing into a paper bag is an example of rebreathing. In this activity, we will observe what happens to pH and carbon dioxide levels in the blood during rebreathing.

1. Click **Start.** Allow the normal breathing trace to run for 10 seconds; then at the 10 second mark, click **Rebreathing.** Watch the pH meter display, as well as the readings in the P_{CO_2} displays and the shape of the trace. As the trace runs, record the readings for pH at each of the following times:

At 20 seconds, pH = ______________

At 40 seconds, pH = ______________

At 60 seconds, pH = ______________

2. Allow the trace to run all the way across the oscilloscope screen and end.
3. Click **Record Data.**
4. If you have printer access, click **Tools** at the top of the screen and select **Print Graph.** Otherwise, manually sketch what you see on the oscilloscope screen.
5. Click **Clear Tracings** to clear the oscilloscope screen.

Did the pH level of the blood change at all during this run? If so, how?

____

Was the pH level always within the "normal" range for the

human body? ______________________________

If not, when was the pH value outside of the normal range, and what acid-base imbalance did this pH value indicate?

____

Did the P_{CO_2} level change during the course of this run? If so, how?

___

If you observed an acid-base imbalance during this run, how would you expect the renal system to compensate for this condition?

___

How did the rebreathing trace differ from the trace for normal breathing? Did the tidal volumes change?

___

___

Give examples of respiratory problems that would result in pH and P_{CO_2} patterns similar to what you observed during rebreathing.

___

___

6. To print out all of the recorded data from this activity, click **Tools** and then **Print Data.** ■

In the next set of activities, we will focus on the body's primary mechanism of compensating for respiratory acidosis or alkalosis: renal compensation.

Renal System Compensation

The kidneys play a major role in maintaining fluid and electrolyte balance in the body's internal environment. By regulating the amount of water lost in the urine, the kidneys defend the body against excessive hydration or dehydration. By regulating the excretion of individual ions, the kidneys maintain normal electrolyte patterns of body fluids. By regulating the acidity of urine and the rate of electrolyte excretion, the kidneys maintain plasma pH levels within normal limits. Renal compensation is the body's primary method of compensating for conditions of respiratory acidosis or respiratory alkalosis. (Although the renal system also compensates for metabolic acidosis or metabolic alkalosis, a more immediate mechanism for compensating for metabolic acid-base imbalances is the respiratory system, as we will see in a later experiment.)

The activities in this section examine how the renal system compensates for respiratory acidosis or alkalosis. The primary variable we will be working with is P_{CO_2} (the partial pressure of carbon dioxide in the blood). We will observe how increases and decreases in P_{CO_2} affect the levels of $[H^+]$ and $[HCO_3^-]$ (bicarbonate) that the kidneys excrete in urine.

Click on **Experiment** at the top of the screen and select **Renal System Compensation.** You will see the screen shown in Figure 10.2. If you completed Exercise 9 on renal physiology, this screen should look familiar. There are two beakers on the left side of the screen, one of which is filled with blood, simulating the body's blood supply to the kidneys. Notice that the P_{CO_2} level is currently set to 40, and that the corresponding pH value is 7.4—both "normal" values. By clicking **Start,** you will initiate the process of delivering blood to the simulated nephron at the right side of the screen. As blood flows through the glomerulus of the nephron, you will see the filtration from the plasma of everything except proteins and cells (note that the moving red dots in the animation do *not* include red blood cells). Blood will then drain from the glomerulus to the beaker at the right of the original beaker. At the end of the nephron tube, you will see the collection of urine in a small beaker. Keep in mind that although only one nephron is depicted here, there are actually over a million nephrons in each human kidney. Below the urine beaker are displays for H^+ and HCO_3^-, which will tell us the relative levels of these ions present in the urine.

Activity 4: Renal Response to Normal Acid-Base Balance

1. Set the P_{CO_2} value to 40, if it is not already. (To increase or decrease P_{CO_2}, click the (−) or (+) buttons. Notice that as P_{CO_2} changes, so does the blood pH level.)
2. Click **Start** and allow the run to finish.
3. At the end of the run, click **Record Data.**

At normal P_{CO_2} and pH levels, what level of H^+ was present in the urine? ___

What level of $[HCO_3^-]$ was present in the urine? ___

Why does the blood pH value change as P_{CO_2} changes?

___

___

4. Click **Refill** to prepare for the next activity. ■

Activity 5: Renal Response to Respiratory Alkalosis

In this activity, we will simulate respiratory alkalosis by setting the P_{CO_2} to values lower than normal (thus, blood pH will be *higher* than normal). We will then observe the renal system's response to these conditions.

1. Set P_{CO_2} to 35 by clicking the (−) button. Notice that the corresponding blood pH value is 7.5.
2. Click **Start.**
3. At the end of the run, click **Record Data.**
4. Click **Refill.**

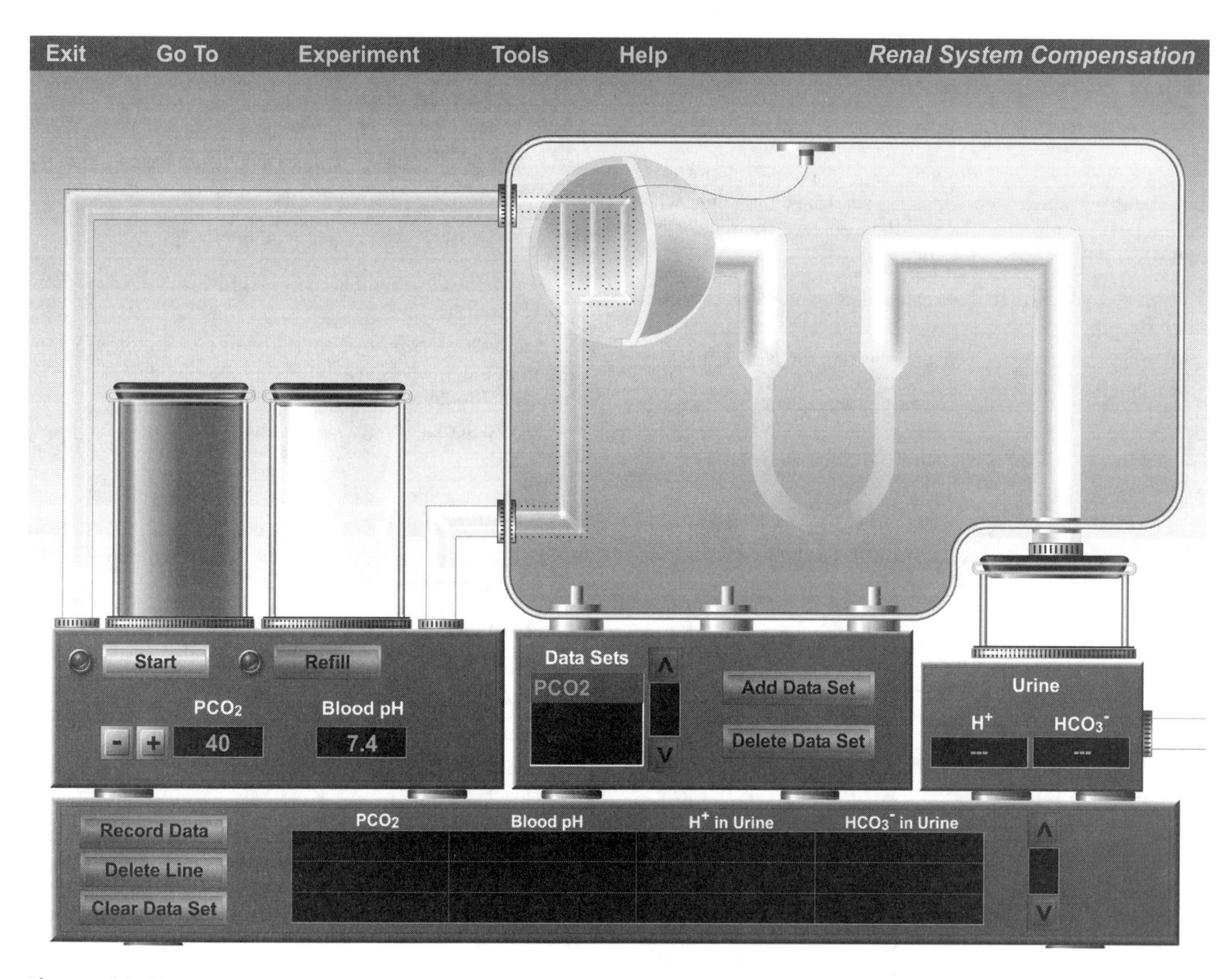

Figure 10.2 Opening screen of the Renal Compensation experiment.

5. Repeat steps 1–4, setting P_{CO_2} to increasingly lower values (i.e., set P_{CO_2} to 30 and then 20, the lowest value allowed).

What level of $[H^+]$ was present in the urine at each of these P_{CO_2}/pH levels?

____

What level of $[HCO_3^-]$ was present in the urine at each of these P_{CO_2}/pH levels?

____

Recall that it may take hours or even days for the renal system to respond to disruptions in acid-base balance. Assuming that enough time has passed for the renal system to fully compensate for respiratory alkalosis, would you expect P_{CO_2} levels to increase or decrease? Would you expect blood pH levels to increase or decrease?

____

____

Recall your activities in the first experiment on respiratory acidosis and alkalosis. Which type of breathing resulted in P_{CO_2} levels closest to the ones we experimented with in this activity—normal breathing, hyperventilation, or rebreathing?

____

Explain why this type of breathing resulted in alkalosis.

____

____ ■

Activity 6:
Renal Response to Respiratory Acidosis

In this activity, we will simulate respiratory acidosis by setting the P_{CO_2} values higher than normal (thus, blood pH will be *lower* than normal). We will then observe the renal system's response to these conditions.

1. Make sure the left beaker is filled with blood. If not, click **Refill.**
2. Set P_{CO_2} to 60 by clicking the (+) button. Notice that the corresponding blood pH value is 7.3.
3. Click **Start.**
4. At the end of the run, click **Record Data.**
5. Click **Refill.**
6. Repeat steps 1–5, setting P_{CO_2} to increasingly higher values (i.e., set P_{CO_2} to 75 and then 90, the highest value allowed).

What level of $[H^+]$ was present in the urine at each of these P_{CO_2}/pH levels?

____

What level of $[HCO_3^-]$ was present in the urine at each of these P_{CO_2}/pH levels?

____

Recall that it may take hours or even days for the renal system to respond to disruptions in acid-base balance. Assuming that enough time has passed for the renal system to fully compensate for respiratory acidosis, would you expect P_{CO_2} levels to increase or decrease? Would you expect blood pH levels to increase or decrease?

____

____

Recall your activities in the first experiment on respiratory acidosis and alkalosis. Which type of breathing resulted in P_{CO_2} levels closest to the ones we experimented with in this activity—normal breathing, hyperventilation, or rebreathing?

____

Explain why this type of breathing resulted in acidosis.

____

____

7. Before going on to the next activity, select **Tools** and then **Print Data** in order to save a hard copy of your data results. ■

Metabolic Acidosis and Alkalosis

Conditions of acidosis or alkalosis that do not have respiratory causes are termed *metabolic acidosis* or *metabolic alkalosis.*

Metabolic acidosis is characterized by low plasma HCO_3^- and pH. The causes of metabolic acidosis include:

- *Ketoacidosis,* a buildup of keto acids that can result from diabetes mellitus
- *Salicylate poisoning,* a toxic condition resulting from ingestion of too much aspirin or oil of wintergreen (a substance often found in laboratories)
- The ingestion of too much alcohol, which metabolizes to acetic acid
- Diarrhea, which results in the loss of bicarbonate with the elimination of intestinal contents
- Strenuous exercise, which may cause a buildup of lactic acid from anaerobic muscle metabolism

Metabolic alkalosis is characterized by elevated plasma HCO_3^- and pH. The causes of metabolic alkalosis include:

- Alkali ingestion, such as antacids or bicarbonate
- Vomiting, which may result in the loss of too much H^+
- Constipation, which may result in reabsorption of elevated levels of HCO_3^-

Increases or decreases in the body's normal metabolic rate may also result in metabolic acidosis or alkalosis. Recall that carbon dioxide—a waste product of metabolism—mixes with water in plasma to form carbonic acid, which in turn forms H^+:

$$H_2O + \mathbf{CO_2} \rightarrow \underset{\text{carbonic acid}}{H_2CO_3} \rightarrow \mathbf{H^+} + \underset{\text{bicarbonate ion}}{HCO_3^-}$$

Therefore, an increase in the normal rate of metabolism would result in more carbon dioxide being formed as a metabolic waste product, resulting in the formation of more H^+—lowering plasma pH and potentially causing acidosis. Other acids that are also normal metabolic waste products, such as ketone bodies and phosphoric, uric, and lactic acids, would likewise accumulate with an increase in metabolic rate. Conversely, a decrease in the normal rate of metabolism would result in less carbon dioxide being formed as a metabolic waste product, resulting in the formation of less H^+—raising plasma pH and potentially causing alkalosis. Many factors can affect the rate of cell metabolism. For example, fever, stress, or the ingestion of food all cause the rate of cell metabolism to increase. Conversely, a fall in body temperature or a decrease in food intake causes the rate of cell metabolism to decrease.

The respiratory system compensates for metabolic acidosis or alkalosis by expelling or retaining carbon dioxide in the blood. During metabolic acidosis, respiration increases to expel carbon dioxide from the blood and decrease $[H^+]$ in order to raise the pH level. During metabolic alkalosis, respiration decreases to promote the accumulation of carbon

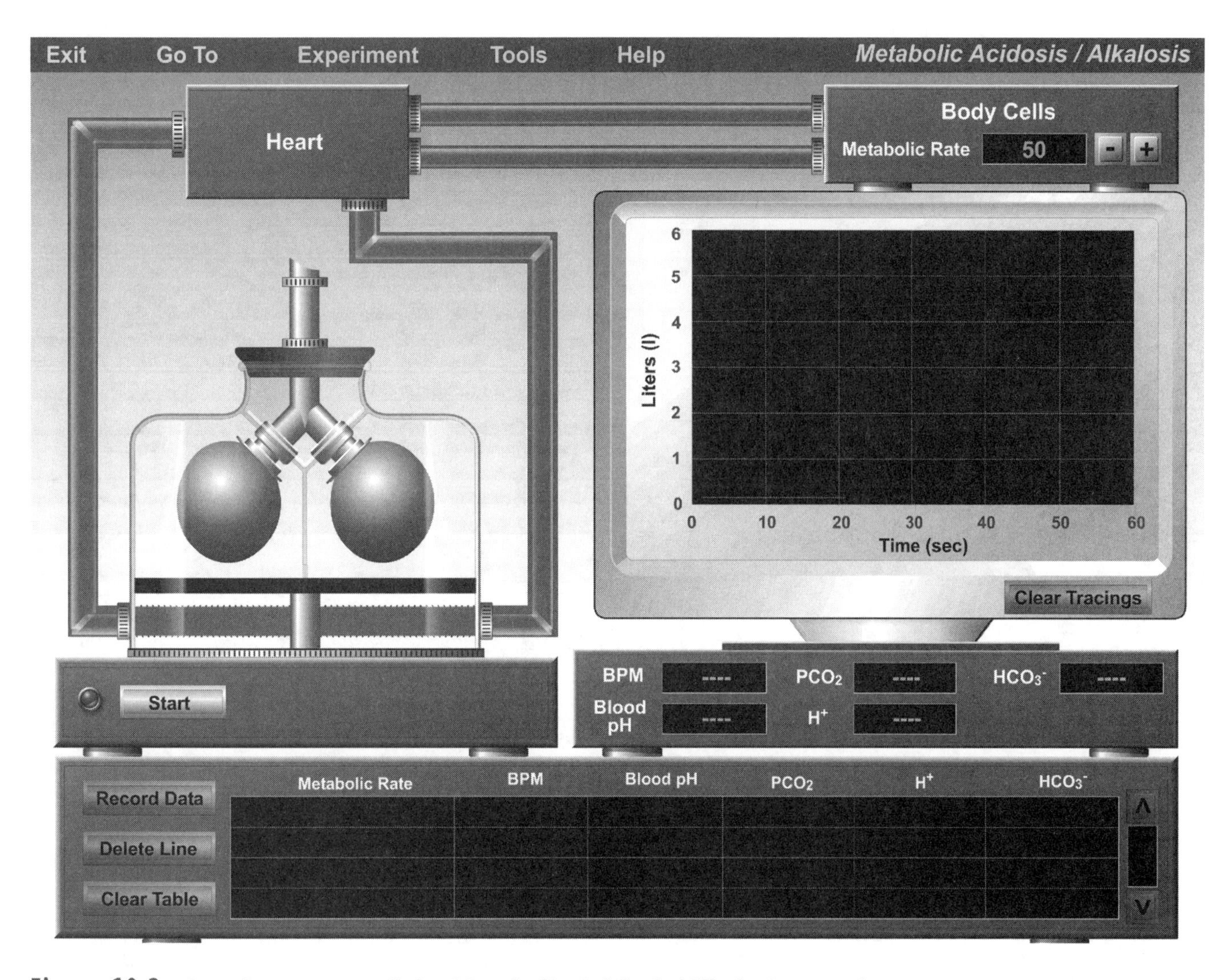

Figure 10.3 Opening screen of the Metabolic Acidosis/Alkalosis experiment.

dioxide in the blood, thus increasing [H^+] and decreasing the pH level.

The renal system also compensates for metabolic acidosis and alkalosis by conserving or excreting bicarbonate ions. However, in this set of activities we will focus on respiratory compensation of metabolic acidosis and alkalosis.

To begin, click **Experiment** at the top of the screen and select **Metabolic Acidosis/Alkalosis.** The screen shown in Figure 10.3 will appear. This screen is similar to the screen from the first experiment; the main differences are the addition of a box representing the heart; tubes showing the double circulation of the heart; and a box representing the body's cells. The default "normal" metabolic rate has been set to 50 kcal/h—an arbitrary value, given that "normal" metabolic rates vary widely from individual to individual. The (+) and (−) buttons in the Body Cells box allow you to increase or decrease the body's metabolic rate. In the following activities, we will observe the respiratory response to acidosis or alkalosis brought on by increases or decreases in the body's metabolic rate.

Activity 7:
Respiratory Response to Normal Metabolism

We will begin by observing respiratory activity at normal metabolic conditions. This data will serve as a baseline against which we will compare our data in Activities 8 and 9.

1. Make sure the Metabolic Rate is set to 50, which for the purposes of this experiment we will consider the "normal" value.

2. Click **Start** to begin the experiment. Notice the arrows showing the direction of blood flow. A graph displaying respiratory activity will appear on the oscilloscope screen.

3. After the graph has reached the end of the screen, the experiment will automatically stop. Note the data in the displays below the oscilloscope screen:

- The **BPM** display gives you the *breaths-per-minute*—the rate at which respiration occurred.

- Blood pH tells you the pH value of the blood.
- PCO_2 (shown as P_{CO_2} in the text) tells you the partial pressure of carbon dioxide in the blood.
- H^+ and HCO_3^- tell you the levels of each of these ions present.

4. Click **Record Data.**
5. Click **Tools** and then **Print Graph** in order to print your graph.

What is the respiratory rate? ____________________

Are the blood pH and P_{CO_2} values within normal ranges?

____

6. Click **Clear Tracings** before proceeding to the next activity. ■

Activity 8: Respiratory Response to Increased Metabolism

1. Increase the metabolic rate to 60.
2. Click **Start** to begin the experiment.
3. Allow the graph to reach the end of the oscilloscope screen. Note the data in the displays below the oscilloscope screen.
4. Click **Record Data.**
5. Click **Tools** and then **Print Graph** in order to print your graph.
6. Repeat steps 1–5 with the metabolic rate set at 70, and then 80.

As the body's metabolic rate increased:

How did respiration change?

____

How did blood pH change?

____

How did P_{CO_2} change?

____

How did $[H^+]$ change?

____

How did $[HCO_3^-]$ change?

____

Explain why these changes took place as metabolic rate increased.

____

____

Which metabolic rates caused pH levels to decrease to a condition of metabolic acidosis?

____

What were the pH values at each of these rates?

____

By the time the respiratory system fully compensated for acidosis, how would you expect the pH values to change?

____

7. Click **Clear Tracings** before proceeding to the next activity. ■

Activity 9: Respiratory Response to Decreased Metabolism

1. Decrease the metabolic rate to 40.
2. Click **Start** to begin the experiment.
3. Allow the graph to reach the end of the oscilloscope screen. Note the data in the displays below the oscilloscope screen.
4. Click **Record Data.**
5. Click **Tools** and then **Print Graph** in order to print your graph.
6. Repeat steps 1–5 with the metabolic rate set at 30, and then 20.

As the body's metabolic rate decreased:

How did respiration change?

____

How did blood pH change?

____

How did P_{CO_2} change?

____

How did $[H^+]$ change?

____

How did [HCO_3^-] change?

____

Explain why these changes took place as the metabolic rate decreased.

____

____

____

Which metabolic rates caused pH levels to increase to a condition of metabolic alkalosis?

____

What were the pH values at each of these rates?

____

By the time the respiratory system fully compensated for alkalosis, how would you expect the pH values to change?

____

7. Click **Tools** → **Print Data** to print your recorded data. ■

Using the Histology Module

Exercise

H

Examining a specimen using a microscope accomplishes two goals: first, it gives you an understanding of the cellular organization of tissues, and second, perhaps even more importantly, it hones your observational skills. Because developing these skills is crucial in the understanding and eventual mastery of the way of thinking in science, using this histology module is not intended as a substitute for using the microscope. Instead, use the histology module to gain an overall appreciation of the specimens and then make your own observations using your microscope. The histology module is also an excellent review tool.

For a review of histology slides specific to topics covered in the PhysioEx lab simulations, turn to the Histology Review Supplement on p. P-141.

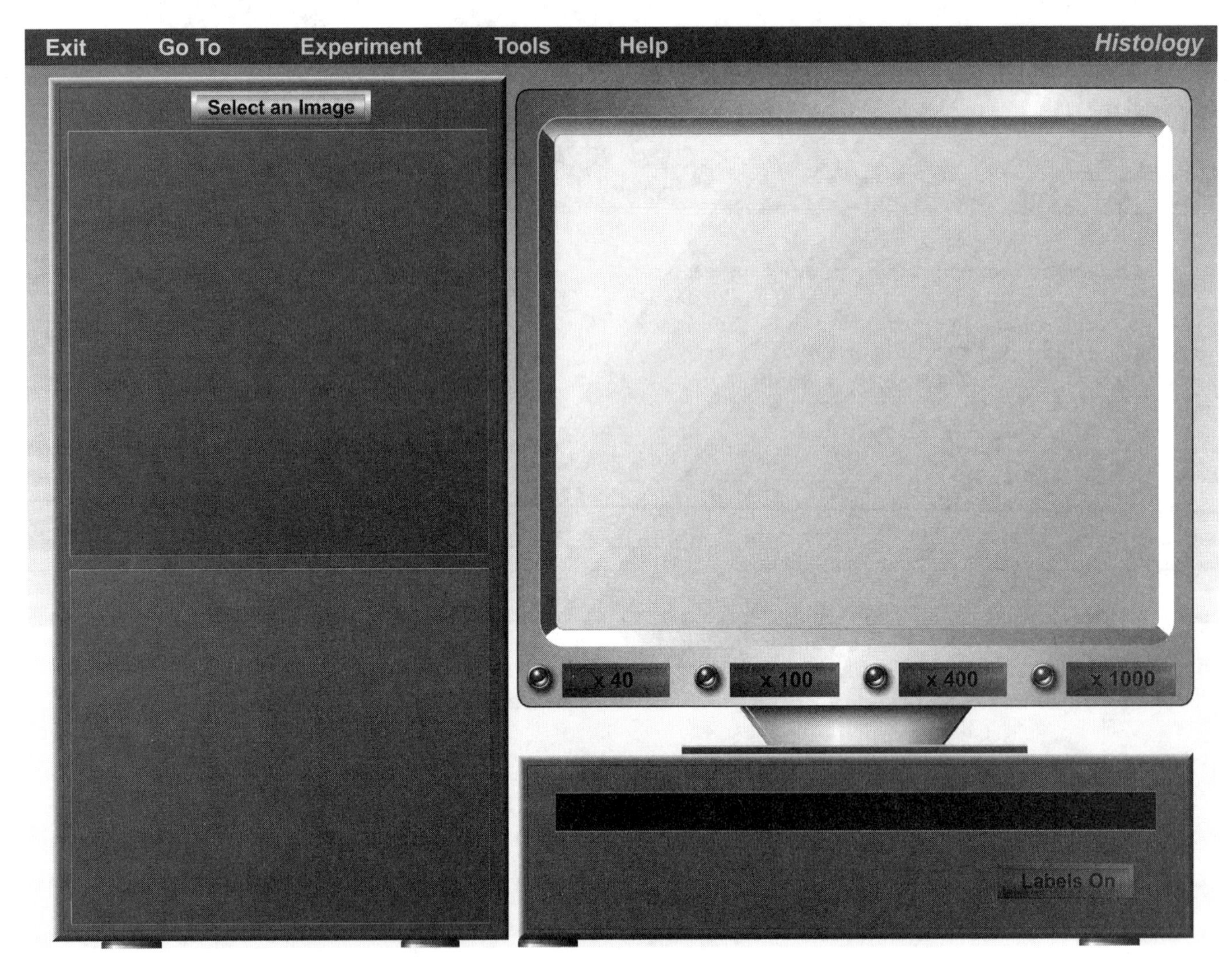

Figure H Opening screen of the Histology Tutorial.

Exploring Digital Histology

Choose **Histology Tutorial** from the main menu. The opening screen for the Histology module will appear in a few seconds (Figure H). The main features on the screen when the program starts are a pair of empty boxes at the left side of the screen, a large image-viewing window on the right side, and a set of magnification buttons below the image window.

Click **Select an Image.** You will see an A-to-Z index appear. Roll your mouse over each letter to see the list of images available. Highlight the image you want to view, and click on it. The image will appear in the image-viewing window.

The text description of the specimen is displayed in the large text box at the upper left and the image title underneath the image-viewing window. Occasionally the words in the text are highlighted in color; clicking those words displays a second image that you can compare with the main image. Click the **Close** button to exit the second image before selecting a new image.

Click and hold the mouse button (left button on a PC) on an image and then drag the mouse to move other areas of the specimen into the image-viewing window, much like moving a slide on a microscope stage.

Choose the magnification of the image in the viewing window by clicking one of the magnification buttons below the image window. Please note that a red magnification button indicates that the slide is not available for viewing at that magnification.

To see parts of the slide labeled, click **Labels On**. Click the button again to remove the labels.

NAME ______________________ LAB TIME/DATE ______________________

REVIEW SHEET
exercise
1

Cell Transport Mechanisms and Permeability: Computer Simulation

Choose all answers that apply to items 1 and 2, and place their letters on the response blanks to the right.

1. Differential permeability: ______________

a. is also called selective permeability
b. refers to the ability of the plasma membrane to select what passes through it
c. implies that all substances pass through membranes without hindrance
d. keeps wastes inside the cell and nutrients outside the cell

2. Passive transport includes: ______________

a. osmosis   b. simple diffusion   c. bulk-phase endocytosis   d. pinocytosis   e. facilitated diffusion

3. The following refer to the dialysis simulation.

Did the 20 MWCO membrane exclude any solute(s)? ______________________________

Which solute(s) passed through the 100 MWCO membrane? ______________________________

Which solute exhibited the highest diffusion rate through the 100 MWCO membrane? ______________________________

Explain why this is so: ______________________________

4. The following refer to the facilitated diffusion simulation.

Are substances able to travel against their concentration gradient? ______________________________

Name two ways to increase the rate of glucose transport. ______________________________

Did NaCl affect glucose transport? ______________________________

Does NaCl require a transport protein for diffusion? ______________________________

5. The following refer to the osmosis simulation.

Does osmosis require energy? ______

Is water excluded by any of the dialysis membranes? ______

Is osmotic pressure generated if solutes freely diffuse? ______

Explain how solute concentration affects osmotic pressure. ______

______

6. The following refer to the filtration simulation.

What does the simulated filtration membrane represent in a living organism? ______

______

What characteristic of a solute determines whether or not it passes through a filtration membrane?

______

Would filtration occur if we equalized the pressure on both sides of a filtration membrane? ______

7. The following questions refer to the active transport simulation.

Does the presence of glucose carrier proteins affect Na^+ transport? ______

Can Na^+ be transported against its concentration gradient? ______

Are Na^+ and K^+ transported in the same direction? ______

The ratio of Na^+ to K^+ transport is ______ Na^+ transported out of the cell for every ______ K^+ transported into the cell.

8. What single characteristic of the semipermeable membranes used in the simple diffusion and filtration experiments determines which substances pass through them? ______

In addition to this characteristic, what other factors influence the passage of substances through living membranes?

______

9. Assume the left beaker contains 4 m*M* NaCl, 9 m*M* glucose, and 10 m*M* albumin. The right beaker contains 10 m*M* NaCl, 10 m*M* glucose, and 40 m*M* albumin. Furthermore, the dialysis membrane is permeable to all substances except albumin. State whether the substance will move (a) to the right beaker, (b) to the left beaker, or (c) not move.

Glucose ______ Albumin ______

Water ______ NaCl ______

10. Assume you are conducting the experiment illustrated below. Both hydrochloric acid (HCl) with a molecular weight of about 36.5 and ammonium hydroxide (NH_4OH) with a molecular weight of 35 are volatile and easily enter the gaseous state. When they meet, the following reaction will occur:

$$HCl + NH_4OH \rightarrow H_2O + NH_4Cl$$

Ammonium chloride (NH_4Cl) will be deposited on the glass tubing as a smoky precipitate where the two gases meet. Predict which gas will diffuse more quickly and indicate to which end of the tube the smoky precipitate will be closer.

a. The faster diffusing gas is ______________________.

b. The precipitate forms closer to the ______________________ end.

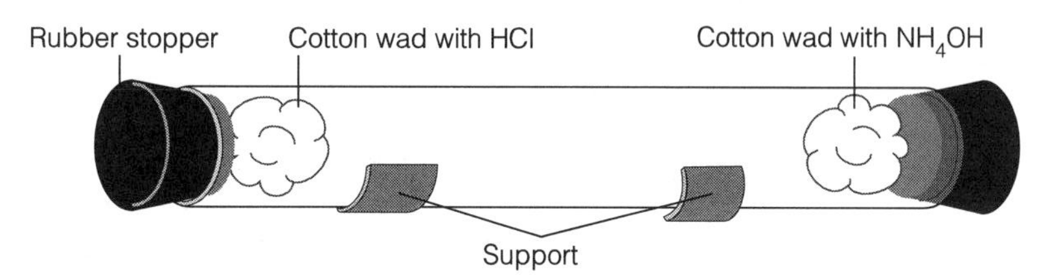

11. When food is pickled for human consumption, as much water as possible is removed from the food. What method is used to achieve this dehydrating effect? ______________________

____

12. What determines whether a transport process is active or passive? ______________________

____

13. Characterize passive and active transport as fully as possible by choosing all the phrases that apply and inserting their letters on the answer blanks.

Passive transport ______________ Active transport ______________

a. accounts for the movement of fats and respiratory gases through the plasma membrane
b. explains solute pumping, bulk-phase endocytosis, and pinocytosis
c. includes osmosis, simple diffusion, and filtration
d. may occur against concentration and/or electrical gradients
e. uses hydrostatic pressure or molecular energy as the driving force
f. moves ions, amino acids, and some sugars across the plasma membrane

14. Define the following:

diffusion: ______________________________

____

osmosis: ______________________________

____

simple diffusion: ______________________________

____

filtration: ______________________________

____

active transport: ______________________________

____

bulk-phase endocytosis: ______________________________

____

pinocytosis: ______________________________

____

facilitated diffusion: ______________________________

____

NAME ______________________________ LAB TIME/DATE ____________________

REVIEW SHEET

exercise

2

Skeletal Muscle Physiology: Computer Simulation

Electrical Stimulation

1. Complete the following statements by filling in your answer on the lines provided below.

A motor unit consists of a __a__ and all the __b__ it innervates. Whole muscle contraction is a(n) __c__ response. In order for muscles to work in a practical sense, __d__ is the method used to produce a slow, steady increase in muscle force.

When we see the slightest evidence of force production on a tracing, the stimulus applied must have reached __e__.

The weakest stimulation that will elicit the strongest contraction that a muscle is capable of is called the __f__. That level of contraction is called the __g__.

When the __h__ of stimulation is so high that the muscle tracing shows fused peaks, __i__ has been achieved.

a. ______________________________ f. ______________________________

b. ______________________________ g. ______________________________

c. ______________________________ h. ______________________________

d. ______________________________ i. ______________________________

e. ______________________________

2. Name each phase of a typical muscle twitch and describe what is happening in each phase.

a. ______________________________

____

b. ______________________________

____

c. ______________________________

____

3. Explain how the PhysioEx experimental muscle stimulation differs from the *in vivo* stimulation via the nervous system. (Note that the graded muscle response following both stimulation methods is similar.)

____

____

____

4. What are the two *experimental* ways in which mode of stimulation can affect the muscle force?

_______________________ and _______________________

Explain your answer.

____

____

____

Isometric Contraction

1. Identify the following conditions by choosing one of the key terms listed on the right.

Key:

_______ is generated by muscle tissue when it is being stretched — a. Total force

_______ requires the input of energy — b. Resting force

_______ is measured by recording instrumentation during contraction — c. Active force

2. Circle the correct response in the parentheses for each statement.

An increase in resting length results in a(n) (increase/decrease) in passive force.

The active force initially (increased/decreased) and then (increased/decreased) as the resting length was increased from minimum to maximum.

As the total force increased, the active force (increased/decreased).

3. Explain what happens to muscle force production at extremes of length (too short or too long). Hint: think about sarcomere structure.

Muscle too short: ____

____

Muscle too long: ____

____

Isotonic Contraction

1. Assuming a fixed starting length, describe the effect resistance has on the initial velocity of shortening, and explain why it has this effect.

____

____

____

2. A muscle has just been stimulated under conditions that will allow both isometric and isotonic contractions. Describe what is happening in terms of length and force.

Isometric: ____

____

Isotonic: ____

____

Terms

Select the condition from column B that most correctly identifies the term in column A.

	Column A		Column B
______	1. muscle twitch	a.	response is all-or-none
______	2. wave summation	b.	affects the force a muscle can generate
______	3. multiple motor unit summation	c.	a single contraction of intact muscle
______	4. resting length	d.	recruitment
______	5. resistance	e.	increasing force produced by increasing stimulus frequency
______	6. initial velocity of shortening	f.	muscle length changing due to relaxation
______	7. isotonic shortening	g.	caused by application of maximal stimulus
______	8. isotonic lengthening	h.	weight
______	9. motor unit	i.	exhibits graded response
______	10. whole muscle	j.	high values with low resistance values
______	11. tetanus	k.	changing muscle length due to active forces
______	12. maximal response	l.	recording shows no evidence of muscle relaxation

NAME ______________________ LAB TIME/DATE ______________________

REVIEW SHEET

exercise

3

Neurophysiology of Nerve Impulses: Computer Simulation

The Nerve Impulse

1. Match each of the terms in column B with the appropriate definition in column A.

Column A	Column B
______ 1. term used to denote that a membrane potential is sitting at about −70 mV	a. threshold
______ 2. reversal of membrane potential due to influx of sodium ions	b. sodium
______ 3. major cation found outside the cell	c. potassium
______ 4. minimal stimulus needed to elicit an action potential	d. resting membrane potential
______ 5. period when cell membrane is totally insensitive to additional stimuli, regardless of the stimulus force used	e. absolute refractory period
______ 6. major cation found inside the cell	f. depolarization

2. Fill in the blanks with the correct words or terms.

Neurons, as with other excitable cells of the body, have two major physiological properties: ______________ and ______________. A neuron has a positive charge on the outer surface of the cell membrane due in part to the action of an active transport system called the ______________. This system moves ______________ out of the cell and ______________ into the cell. The inside of the cell membrane will be negative, not only due to the active transport system but also because of ______________, which remain negative due to intracellular pH and keep the inside of the cell membrane negative.

3. Why don't the terms *depolarization* and *action potential* mean the same thing?

____

____

____

4. What is the difference between membrane irritability and membrane conductivity?

____

____

____

Eliciting a Nerve Impulse

1. Why does the nerve action potential increase slightly when you add 1.0 V to the threshold voltage and stimulate the nerve?

2. If you were to spend a lot of time studying nerve physiology in the laboratory, what type of stimulus would you use and why?

3. Why does the addition of sodium chloride elicit an action potential?

Inhibiting a Nerve Impulse

1. What was the effect of ether on eliciting an action potential?

2. Does the addition of ether to the nerve cause any permanent alteration in neural response?

3. What was the effect of curare on eliciting an action potential?

4. Explain the reason for your answer to question 3 above.

5. What was the effect of lidocaine on eliciting an action potential?

Nerve Conduction Velocity

1. What is the relationship between size of the nerve and conduction velocity? ______________________

____

____

2. Keeping your answer to question 1 in mind, how might you draw an analogy between the nerves in the human body and electrical wires?

____

3. Hypothesize what types of animals would have the fastest conduction velocities.

____

4. How does myelination affect nerve conduction velocity? Explain. ______________________

____

____

____

5. If any of the nerves used were reversed in their placement on the stimulating and recording electrodes, would any differences be seen in conduction velocity? Explain.

____

____

____

NAME ______________________ LAB TIME/DATE ______________________

REVIEW SHEET

exercise

4

Endocrine System Physiology: Computer Simulation

1. In the following columns, match the hormone on the left with its source on the right.

________ thyroxine	a. ovary
________ estrogen	b. thyroid gland
________ thyroid-stimulating hormone (TSH)	c. pancreas
________ insulin	d. pituitary gland

2. Each hormone is known to have a specific target tissue. For each hormone listed, what is its target tissue and what is its specific action?

thyroxine ______________________

estrogen ______________________

thyroid-stimulating hormone (TSH) ______________________

insulin ______________________

follicle-stimulating hormone (FSH) ______________________

____

Metabolism

1. In the metabolism experiment, what was the effect of thyroxine on the overall metabolic rate of the animals?

____

____

2. Using the respirometer-manometer, you observed the amount of oxygen being used by animals in a closed chamber. What happened to the carbon dioxide the animals produced while in the chamber?

____

____

3. If the experimental animals in the chamber were engaged in physical activity (such as running in a wheel), how do you think this would change the results of the metabolism experiment?

____

What changes would you expect to see in fluid levels of the manometer? ______________________________

____

4. Why didn't the administration of thyroid-stimulating hormone (TSH) have any effect on the metabolic rate of the thyroidectomized rat? ____

____

5. Why didn't the administration of propylthiouracil have any effect on the metabolic rate of either the thyroidectomized rat or the hypophysectomized rat? ____

____

Hormone Replacement Therapy

1. In the experiment with hormone replacement therapy, what was the effect of removing the ovaries from the animals?

____

____

2. Specifically, what hormone did the ovariectomies effectively remove from the animals, and what purpose does this hormone serve? ____

____

3. If a hormone such as testosterone were used in place of estrogen, would any effect be seen? Explain your answer.

____

____

4. In the experiment, you administered 7 injections of estrogen to the experimental rat over the course of 7 days. What do you think would happen if you administered one injection of estrogen per day for an additional week?

____

____

5. What do you think would happen if you administered 7 injections of estrogen to the experimental rat all in one day?

____

6. In a wet lab, why would you need to wait several weeks after the animals received their ovariectomies before you could perform this experiment on them? ____

____

Insulin and Diabetes

1. In the insulin and diabetes experiment, what was the effect of administering alloxan to the experimental animal?

____

2. When insulin travels to the cells of the body, the concentration of what compound will elevate within the cells?

____

What is the specific action of this compound, within the cells? ____

____

3. Name the diseases described below:

   a. the condition when insulin is not produced by the pancreas: ____

   b. the condition when insulin is produced by the pancreas, but the body fails to respond to the insulin:

   ____

4. What was the effect of administering insulin to the diabetic rat? ____

____

5. What is a *glucose standard curve,* and why did you need to obtain one for this experiment? ____

____

6. Would altering the light wavelength of the spectrophotometer have any bearing on the results obtained? Explain your answer. ____

____

7. What would you do to help a friend who had inadvertently taken an overdose of insulin? Why? ____

____

NAME ______________________ LAB TIME/DATE ______________________

REVIEW SHEET

exercise

5

Cardiovascular Dynamics: Computer Simulation

For numbers 1 and 2 below, choose all answers that apply and place their letters on the response blanks to the right of the statement.

1. The circulation of blood through the vascular system is influenced by ______________.

a. blood viscosity
b. the length of blood vessels
c. the driving pressure behind the blood
d. the radius or diameter of blood vessels

2. Peripheral resistance depends on ______________.

a. blood viscosity
b. blood pressure
c. vessel length
d. vessel radius

3. Complete the following statements.

The volume of blood remaining in the heart after ventricular contraction is called the ______________________.

Cardiac output is defined as ____.

The amount of blood pumped by the heart in a single beat is called the ______________________ volume.

The human heart is actually two individual pumps working in ______________________.

Stroke volume is calculated by ____.

If stroke volume decreased, heart rate would ______________________ in order to maintain blood flow.

4. How could the heart compensate to maintain proper blood flow for the following conditions?

High peripheral resistance: ____

A leaky atrioventricular valve: ____

A constricted semilunar valve: ____

5. How does the size of the heart change under conditions of chronic high peripheral resistance?

____

6. The following questions refer to the Vessel Resistance experiment.

How was the flow rate affected when the radius of the flow tube was increased? ______________________

Which of the adjustable parameters had the greatest effect on fluid flow? ______________________

How does vessel length affect fluid flow? ______________________

If you increased fluid viscosity, what parameter(s) could you adjust to keep fluid flow constant?

____

Explain your answer. ____

____

If the driving pressure in the left beaker was 100 mm Hg, how could you adjust the conditions of the experiment to completely stop fluid flow? ____

7. The following questions refer to the Pump Mechanics experiment.

What would happen if the right side of the heart pumped faster than the left side of the heart?

____

____

When you change the radius of the right flow tube in Pump Mechanics, the resulting plot looks different than the radius plot in the Vessel Resistance experiment. How would the plot look if you changed the radius of both flow tubes in Pump Mechanics instead of just the right flow tube?

____

____

Why are valves needed in the Pump Mechanics equipment? ____

What happens to blood flow if peripheral resistance equals pump pressure? ____

Theoretically, what would happen to the pumping ability of the heart if the end systolic volume was equal to the end diastolic volume? ____

8. Match the part in the simulation equipment to the analogous cardiac structure or physiological term listed in the key below.

Simulation equipment:	Key:
________ 1. valve leading to the right beaker	a. pulmonary veins
________ 2. valve leading to the pump	b. bicuspid valve
________ 3. left flow tube	c. ventricular filling pressure
________ 4. right flow tube	d. peripheral resistance
________ 5. pump end volume	e. aortic valve
________ 6. pump starting volume	f. end diastolic volume
________ 7. pressure in the right beaker	g. aorta
________ 8. pressure in the left beaker	h. end systolic volume
________ 9. pump pressure	i. systolic pressure

9. Define the following terms:

blood flow: ____

____

peripheral resistance: ____

____

viscosity: ____

____

radius: ____

____

end diastolic volume: ____

____

systole: ____

____

diastole: ____

____

NAME ______________________ LAB TIME/DATE ______________________

REVIEW SHEET
exercise
6

Frog Cardiovascular Physiology: Computer Simulation

Special Electrical Properties of Cardiac Muscle: Automaticity and Rhythmicity

1. Define the following terms:

   Automaticity ______________________

   Rhythmicity ______________________

2. Explain the anatomical differences between frog and human hearts.

   ____

   ____

   ____

Baseline Frog Heart Activity

1. Define the following terms:

   Intrinsic heart control ______________________

   Extrinsic heart control ______________________

2. Why is it necessary to keep the frog heart moistened with Ringer's solution? ______________________

   ____

Refractory Period of Cardiac Muscle

1. Define *extrasystole* ______________________

   ____

2. Refer to the exercise to answer the following questions.

   What was the effect of stimulating the heart during ventricular contraction? ______________________

   During ventricular relaxation ______________________

   During the pause interval ______________________

   What does this information indicate about the refractory period of cardiac muscle?

   ____

   Can cardiac muscle be tetanized? __________ Why or why not? ______________________

The Effect of Vagus Nerve Stimulation

1. What was the effect of vagal stimulation on heart rate? ______________________.

2. What is vagal escape? ______________________

3. Why is the vagal escape valuable in maintaining homeostasis? ______________________

____

Physical and Chemical Modifiers of Heart Rate

1. Describe the effect of thermal factors on the frog heart.

   Cold ______________________ Heat ______________________

2. Which of the following factors caused the same, or very similar, heart rate-reducing effects: epinephrine, atropine, pilocarpine, digitalis, potassium ions.

   ____

   ____

   Which of the factors listed above would reverse or antagonize vagal effects? ______________________

   ____

3. Did administering any of the following produce any changes in force of contraction (shown by peaks of increasing or decreasing height)? If so, explain the mechanism.

   Epinephrine ______________________

   ____

   Calcium ions ______________________

   ____

4. Excessive amounts of each of the following ions would most likely interfere with normal heart activity. Explain the type of changes caused in each case.

   K^+ ______________________

   Ca^{2+} ______________________

   Na^+ ______________________

5. Define the following:

   Parasympathomimetic ______________________

   Ectopic pacemaker ______________________

6. Explain how digitalis works. ______________________

   ____

NAME ______________________ LAB TIME/DATE ______________________

REVIEW SHEET
exercise
7

Respiratory System Mechanics: Computer Simulation

Define the following terms:

1. Ventilation ______________________

2. Inspiration ______________________

3. Expiration ______________________

Measuring Respiratory Volumes

1. Write the respiratory volume term and the normal value that is described by the following statements:

Volume of air present in the lungs after a forceful expiration ______________________

Volume of air that can be expired forcefully after a normal expiration ______________________

Volume of air that is breathed in and out during a normal respiration ______________________

Volume of air that can be inspired forcefully after a normal inspiration ______________________

Volume of air corresponding to TV + IRV + ERV ______________________

2. Fill in the formula for minute respiratory volume:

____

Examining the Effect of Changing Airway Resistance on Respiratory Volumes

1. Even though pulmonary function tests are not diagnostic, they can help determine the difference between

______________ and ______________ disorders.

2. Chronic bronchitis and asthma are examples of ______________ disorders.

3. Describe FEV_1: ______________________

____

4. Explain the difference between FVC and FEV_1: ______________________

____

5. What effect would increasing airway resistance have on FEV_1? ______________________

Examining the Effect of Surfactant

1. Explain the term *surface tension*. ______________________________

____

2. Surfactant is a detergent-like ______________________.

3. How does surfactant work? ______________________________

4. What might happen to ventilation if the watery film lining the alveoli did not contain surfactant?

____

Investigating Intrapleural Pressure

1. Complete the following statements.

The pressure within the pleural cavity, __1__, is __2__ than the pressure within the alveoli. This __3__ pressure condition is caused by two forces, the tendency of the lung to recoil due to its __4__ properties and the __5__ of the alveolar fluid. These two forces act to pull the lungs away from the thoracic wall, creating a partial __6__ in the pleural cavity. Because the pressure in the __7__ space is lower than __8__, any opening created in the thoracic wall equalizes the intrapleural pressure with the atmospheric pressure, allowing air to enter the pleural cavity, a condition called __9__. Pneumothorax allows __10__, a condition called __11__.

1. ______________________
2. ______________________
3. ______________________
4. ______________________
5. ______________________
6. ______________________
7. ______________________
8. ______________________
9. ______________________
10. ______________________
11. ______________________

2. Why is the intrapleural pressure negative rather than positive? ______________________

____

____

3. Would intrapleural pressure be positive or negative when blowing up a balloon? Explain your answer.

____

____

Exploring Various Breathing Patterns

1. Match the term listed in column B with the descriptive phrase in column A. (There may be more than one correct answer.)

Column A	Column B
________1. causes a drop in carbon dioxide concentration in the blood	a. rebreathing
________2. results in lower blood pH	b. rapid breathing
________3. stimulates an increased respiratory rate	c. breath holding
________4. results in a lower respiratory rate	
________5. can be considered an extreme form of rebreathing	
________6. causes a rise in blood carbon dioxide	

2. Because carbon dioxide is the main stimulus for respirations, what would happen to respiratory drive if you held your breath? ____

____

NAME ______________________ LAB TIME/DATE ______________________

REVIEW SHEET
exercise
8

Chemical and Physical Processes of Digestion: Compter Simulation

Chemical Digestion of Foodstuffs: Enzymatic Action

1. Match the following definitions with the proper choices from the key.

Key: a. catalyst b. control c. enzyme d. substrate

_______ 1. increases the rate of a chemical reaction without becoming part of the product

_______ 2. provides a standard of comparison for test results

_______ 3. biological catalyst: protein in nature

_______ 4. substance on which an enzyme works

2. Name three characteristics of enzymes. ______________________

____

3. Explain the following statement: The enzymes of the digestive system are classified as hydrolases.

____

____

4. Fill in the chart below with what you have learned about the various digestive system enzymes encountered in this exercise.

Enzyme	Organ producing it	Site of action	Substrate(s)	Optimal pH
Salivary amylase				
Pepsin				
Lipase (pancreatic)				

5. Name the end products of digestion for the following types of foods.

Proteins: ______________________ Carbohydrates: ______________________

Fats: ______________________ and ______________________

6. You used several different indicators or tests in the laboratory to determine the presence or absence of certain substances. Choose the correct test or indicator from the key to correspond to the condition described below.

Key: a. IKI b. Benedict's solution c. pH meter d. BAPNA

________ 1. used to test for protein hydrolysis, which was indicated by a yellow color

________ 2. used to test for the presence of starch, which was indicated by a blue-black color

________ 3. used to test for the presence of fatty acids

________ 4. used to test for the presence of maltose, which was indicated by a blue to green (or to rust) color change

7. The three-dimensional structure of a functional protein is altered by intense heat or nonphysiological pH even though peptide bonds may not break. Such a change in protein structure is called denaturation, and denatured enzymes are not functional. Explain why.

____

____

8. What experimental conditions in the simulation resulted in denatured enzymes? ______________________________

____

9. Complete the mechanism of absorption section in the chart below for each of the substances listed. Use a check mark to indicate whether the absorption would result in the movement of a substance into the blood capillaries or the lymph capillaries (lacteals).

Substance	**Mechanism of absorption**	**Blood**	**Lymph**
Monosaccharides			
Fatty acids and glycerol			
Amino acids			

10. Imagine that you have been chewing a piece of bread for 5 to 6 minutes. How would you expect its taste to change during this time? ____

11. People on a strict diet to lose weight begin to metabolize stored fats at an accelerated rate. How could this condition affect blood pH? ____

Starch Digestion by Salivary Amylase

1. What conclusions can you draw when an experimental sample gives both a positive starch test and a positive maltose test?

2. Why was 37°C the optimal incubation temperature?

3. Why did very little, if any, starch digestion occur in tube 1?

4. Why did very little starch digestion occur in tubes 6 and 7?

5. Imagine that you have told a group of your peers that amylase is capable of digesting starch to maltose. If you had not run the experiment in control tubes 3, 4, and 5, what objections to your statement could be raised?

Assessing Plant Starch Digestion

1. What is the effect of freezing on enzyme activity?

2. What can you conclude about the ability of salivary amylase to digest plant starch?

3. What can you conclude about the ability of bacteria to digest plant starch?

Protein Digestion by Pepsin

1. Why is an indicator reagent such as IKI or Benedict's solution not necessary when using a substrate like BAPNA?

2. Trypsin is a pancreatic hydrolase present in the small intestine during digestion. Would trypsin work well in the stomach? Explain your answer.

3. How does the optical density of a solution containing BAPNA relate to enzyme activity?

4. What happens to pepsin activity as it reaches the small intestine?

Fat Digestion by Pancreatic Lipase and the Action of Bile

1. Why does the pH of a fatty solution decrease as enzymatic hydrolysis increases? ______________________________

____

2. How does bile affect fat digestion? ______________________________

3. Why is it not possible to determine the activity of lipase in the pH 2.0 buffer using the pH meter assay method?

____

4. Why is bile not considered an enzyme? ______________________________

Physical Processes: Mechanisms of Food Propulsion and Mixing

Complete the following statements. Write your answers in the numbered spaces below.

Swallowing, or __1__, occurs in two phases—the __2__ and __3__. One of these phases, the __4__ phase, is voluntary. During the voluntary phase, the __5__ is used to push the food into the back of the throat. During swallowing, the __6__ rises to ensure that its passageway is covered by the epiglottis so that the ingested substances do not enter the respiratory passageways. It is possible to swallow water while standing on your head because the water is carried along the esophagus involuntarily by the process of __7__. The pressure exerted by the foodstuffs on the __8__ sphincter causes it to open, allowing the food to enter the stomach.

The two major types of propulsive movements that occur in the small intestine are __9__ and __10__. One of these movements, __11__, acts to continually mix the foods and to increase the absorption rate by moving different parts of the chyme mass over the intestinal mucosa, but it has less of a role in moving foods along the digestive tract.

1. ______________________ 7. ______________________

2. ______________________ 8. ______________________

3. ______________________ 9. ______________________

4. ______________________ 10. ______________________

5. ______________________ 11. ______________________

6. ______________________

NAME ____________________ LAB TIME/DATE ____________________

REVIEW SHEET
exercise
9

Renal Physiology—The Function of the Nephron: Computer Simulation

Define the following terms:

1. Glomerulus ____________________

2. Renal tubule ____________________

3. Glomerular capsule ____________________

4. Renal corpuscle ____________________

5. Afferent arteriole ____________________

6. Efferent arteriole ____________________

Investigating the Effect of Flow Tube Radius on Glomerular Filtration

1. In terms of the blood supply to and from the glomerulus, explain why the glomerular capillary bed is unusual.

____

____

2. How would pressure in the glomerulus be affected by constricting the afferent arteriole? Explain your answer.

____

____

3. How would pressure in the glomerulus be affected by constricting the efferent arteriole? Explain your answer.

____

____

Assessing Combined Effects on Glomerular Filtration

1. If systemic blood pressure started to rise, what could the arterioles of the glomerulus do to keep glomerular filtration rate constant? ____________________

____

2. One of the experiments you performed in the simulation was to close the valve at the end of the collecting duct. Is closing that valve more like constricting an afferent arteriole or more like a kidney stone? Explain your answer.

____

____

3. Constricting the efferent arteriole would have the same effect on glomerular filtration as (constricting/dilating) the afferent arteriole.

____

____

Exploring the Role of the Solute Gradient on Maximum Urine Concentration Achievable

Complete the following statements.

In the process of urine formation, solutes and water move from the __1__ of the nephron into the __2__ spaces. The passive movement of solutes and water from the lumen of the renal tubule into the interstitial spaces relies in part on the __3__ surrounding the nephron. When the nephron is permeable to solutes or water, and __4__ will be reached between the interstitial fluid and the contents of the nephron. __5__ is a hormone that increases the water permeability of the __6__ and the collecting duct, allowing water to flow to areas of higher solute concentration, usually from the lumen of the nephron into the surrounding interstitial area. __7__ is hormone that causes __8__ reabsorption at the expense of __9__ loss into the lumen of the tubule.

1. ______________________
2. ______________________
3. ______________________
4. ______________________
5. ______________________
6. ______________________
7. ______________________
8. ______________________
9. ______________________

10. Would the passive movement of substances occur if the interstitial solute concentration was the same as the filtrate solute concentration? Explain your answer.

____

____

Studying the Effect of Glucose Carrier Proteins on Glucose Reabsorption

1. In terms of the function of the nephron, explain why one might find glucose in the urine exiting the collecting duct.

____

____

2. Imagine this scenario: a person has the normal number of glucose carriers in the nephrons yet has glucose in the urine. What could be the cause of this condition? (Hint: think about filtration rate.)

____

____

Testing the Effect of Hormones on Urine Formation

1. Complete the following statements.

The concentration of the __1__ excreted by our kidneys changes depending on our immediate needs. For example, if a person consumes a large quantity of water, the excess water will be eliminated, producing __2__ urine. On the other hand, under conditions of dehydration, there is a clear benefit to being able to produce urine as __3__ as possible, thereby retaining precious water. Although the medullary gradient makes it possible to excrete concentrated urine, urine dilution or concentration is ultimately under __4__ control. In this experiment, you will investigate the effects of two different hormones on renal function, aldosterone produced by the __5__ and ADH manufactured by the __6__ and stored in the __7__. Aldosterone works to reabsorb __8__ (and thereby water) at the expense of losing __9__. Its site of action is the __10__. ADH makes the distal tubule and collecting duct more permeable to __11__, thereby allowing the body to reabsorb more water from the filtrate when it is present.

1. ______________________
2. ______________________
3. ______________________
4. ______________________
5. ______________________
6. ______________________
7. ______________________
8. ______________________
9. ______________________
10. ______________________
11. ______________________

2. Match the term listed in column B with the descriptive phrase in column A. (There may be more than one correct answer.)

Column A	Column B
_______1. causes production of dilute urine	a. increased ADH
_______2. results in increased sodium loss	b. increased aldosterone
_______3. causes the body to retain more potassium	c. decreased ADH
_______4. will cause water retention due to sodium movement	d. decreased aldosterone
_______5. causes water reabsorption due to increased membrane permeability	
_______6. increases sodium reabsorption	

NAME ____________________ LAB TIME/DATE ____________________

REVIEW SHEET

exercise **10**

Acid-Base Balance: Computer Simulation

1. Match each of the terms in column A with the appropriate description in column B.

Column A	Column B
______ 1. pH	a. condition in which the human body's pH levels fall below 7.35
______ 2. acid	b. condition in which the human body's pH levels rise above 7.45
______ 3. base	c. mixes with water in the blood to form carbonic acid
______ 4. acidosis	d. substance that binds to H^+ in solution
______ 5. alkalosis	e. substance that releases H^+ in solution
______ 6. carbon dioxide	f. term used to denote hydrogen ion concentration in body fluids

2. What is the normal range of pH levels of blood and tissue fluids in the human body?

____

3. What is the difference between a *strong acid* and a *weak acid*?

____

4. What is the difference between a *strong base* and a *weak base*?

____

5. What is the difference between respiratory acidosis/alkalosis and metabolic acidosis/alkalosis?

____

____

6. What are the body's two major physiological buffer systems for compensating for acid-base imbalances?

____

Respiratory Acidosis and Alkalosis

1. What are some of the causes of respiratory acidosis?

____

2. What are some of the causes of respiratory alkalosis?

____

3. What happens to blood pH levels during hyperventilation? Why?

4. What happens to blood pH levels during rebreathing? Why?

5. Circle the correct bolfaced terms:

As respiration increases, P_{CO_2} levels **increase / decrease** and pH levels **rise / fall.**

As respiration decreases, P_{CO_2} levels **increase / decrease** and pH levels **rise / fall.**

Renal Compensation

1. How does the renal system compensate for conditions of respiratory acidosis?

2. How does the renal system compensate for conditions of respiratory alkalosis?

Metabolic Acidosis and Alkalosis

1. What are some of the causes of metabolic acidosis?

2. What are some of the causes of metabolic alkalosis?

3. Explain how the respiratory system compensates for metabolic acidosis and alkalosis.

4. Explain how the renal system compensates for metabolic acidosis and alkalosis.

5. Circle the correct bolfaced terms:

As metabolic rate increases, respiration **increases / decreases,** P_{CO_2} levels **increase / decrease,** and pH levels **rise / fall.**

As metabolic rate decreases, respiration **increases / decreases,** P_{CO_2} levels **increase / decrease,** and pH levels **rise / fall.**

Histology Review Supplement

The slides in this section are designed to provide a basic histology review related to topics introduced in the PhysioEx lab simulations and in your anatomy and physiology textbook.

From the PhysioEx main menu, select **Histology Review Supplement.** When the screen comes up, click **Select an Image Group.** You will note that the slides in the histology module are grouped in the following categories:

Skeletal muscle slides

Nervous tissue slides

Endocrine tissue slides

Cardiovascular tissue slides

Respiratory tissue slides

Digestive tissue slides

Renal tissue slides

Select the group of slides you wish to view, and then refer to the relevant worksheet in this section for a step-by-step tutorial. For example, if you would like to review the skeletal muscle slides, click on **Skeletal muscle slides** and then turn to the next page of this lab manual for the worksheet entitled Skeletal Muscle Tissue Review to begin your review.

Since the slides in this module have been selected for their relevance to topics covered in the PhysioEx lab simulations, it is recommended that you complete the worksheets along with a related PhysioEx lab. For example, you might complete the Skeletal Muscle Tissue worksheet right before or after your instructor assigns you Exercise 2, the PhysioEx lab simulation on Skeletal Muscle Physiology.

For additional histology review, turn to page P-103.

Skeletal Muscle Tissue Review

From the PhysioEx main menu, select **Histology Review Supplement.** When the screen comes up, click **Select an Image Group**. From Group Listing, click **Skeletal Muscle Slides**. To view slides without labels, click the **Labels Off** button at the bottom right of the monitor.

Click slide 1.
Skeletal muscle is composed of extremely large, cylindrical multinucleated cells called **myofibers.** The nuclei of the skeletal muscle cell **(myonuclei)** are located peripherally just subjacent to the muscle cell plasmalemma (sarcolemma). The interior of the cell is literally filled with an assembly of contractile proteins (myofilaments) arranged in a specific overlapping pattern oriented parallel to the long axis of the cell.

Click slides 2, 3.
Sarcomeres are the functional units of skeletal muscle. The organization of contractile proteins into a regular end-to-end repeating pattern of sarcomeres along the length of each cell accounts for the striated or striped appearance of skeletal muscle in longitudinal section.

Click slide 4.
The smooth endoplasmic reticulum (sarcoplasmic reticulum), modified into an extensive network of membranous channels that store, release, and take up the calcium necessary for contraction, also functions to further organize the myofilaments inside the cell into cylindrical bundles called myofibrils. The **stippled appearance of the cytoplasm** in cells cut in cross section represents the internal organization of myofilaments bundled into myofibrils by the membranous sarcoplasmic reticulum.

What is the functional unit of contraction in skeletal muscle?

____

What are the two principal contractile proteins that compose the functional unit of contraction?

____

What is the specific relationship of the functional unit of contraction to the striated appearance of a skeletal muscle fiber?

____

____

Click slide 5.
The neural stimulus for contraction arises from the **axon** of a motor neuron whose axon terminal comes into close apposition to the muscle cell sarcolemma.

Would you characterize skeletal muscle as voluntary or involuntary?

____

Name the site of close juxtaposition of an axon terminal with the muscle cell plasmalemma.

____

Skeletal muscle also has an extensive connective tissue component that, in addition to conducting blood vessels and nerves, becomes continuous with the connective tissue of its tendon. The tendon in turn is directly continuous with the connective tissue covering (the periosteum) of the adjacent bone. This connective tissue continuity from muscle to tendon to bone is the basis for movement of the musculoskeletal system.

What is the name of the loose areolar connective tissue covering of an individual muscle fiber?

____

The perimysium is a collagenous connective tissue layer that groups several muscle fibers together into bundles called

______________________________.

Which connective tissue layer surrounds the entire muscle and merges with the connective tissue of tendons and aponeuroses?

____

Nervous Tissue Review

From the PhysioEx main menu, select **Histology Review Supplement.** When the screen comes up, click **Select an Image Group.** From Group Listing, click **Nervous Tissue Slides.** To view slides without labels, click the **Labels Off** button at the bottom right of the monitor.

Nervous tissue is composed of nerve cells (neurons) and a variety of support cells.

Click slide 1.
Each nerve cell consists of a **cell body** (perikaryon) and one or more **cellular processes** (axon and dendrites) extending from it. The cell body contains the **nucleus,** which is typically pale-staining and round or spherical in shape, and the usual assortment of cytoplasmic organelles. Characteristically, the nucleus features a prominent **nucleolus** often described as resembling the pupil of a bird's eye (**"bird's eye"** or "owl's eye" nucleolus).

Click slide 2.
The cytoplasm of the cell body is most often granular in appearance due to the presence of darkly stained clumps of ribosomes and rough endoplasmic reticulum (**Nissl bodies/ Nissl substance).** Generally, a single axon arises from the **cell body** at a pale-staining region (axon hillock), devoid of Nissl bodies. The location and number of dendrites arising from the cell body varies greatly.

Axons and dendrites are grouped together in the peripheral nervous system (PNS) to form peripheral nerves.

What is the primary unit of function in nervous tissue?

____

Name the pale-staining region of the cell body from which the axon arises.

____

The support cells of the nervous system perform extremely important functions including support, protection, insulation, and maintenance and regulation of the microenvironment that surrounds the nerve cells.

Click slides 3, 4.
In the PNS, support cells surround both cell bodies (satellite cells) and individual **axons** and **dendrites** (Schwann cells). Schwann cells, in particular, are responsible for wrapping their cell membrane jelly-roll style around axons and dendrites to form an insulating sleeve called the **myelin sheath.**

Click slide 5.
Because Schwann cells are aligned in series and myelinate only a small segment of a single axon, small gaps occur between the myelin sheaths of adjacent contiguous Schwann cells. The gaps, called **nodes of Ranvier,** together with the insulating properties of **myelin,** enhance the speed of conduction of electrical impulses along the length of the axon. Different support cells and myelinating cells are present in the CNS.

What is the general name for all support cells within the CNS?

____

Name the specific myelinating cell of the CNS.

____

In the PNS, connective tissue also plays a role in providing support and organization. In fact, the composition and organization of the connective tissue investments of peripheral nerves are similar to those of skeletal muscle.

Click slide 3.
Each individual axon or dendrite is surrounded by a thin and delicate layer of loose connective tissue called the endoneurium (not shown.) The **perineurium,** a slightly thicker layer of loose connective tissue, groups many axons and dendrites together into bundles (fascicles). The outermost **epineurium** surrounds the entire nerve with a thick layer of dense irregular connective tissue, often infiltrated with adipose tissue, that conveys blood and lymphatic vessels to the nerve. There is no connective tissue component within the nervous tissue of the CNS.

What is the relationship of the endoneurium to the myelin sheath?

____

____

Endocrine Tissue Review

From the PhysioEx main menu, select **Histology Review Supplement.** When the screen comes up, click **Select an Image Group.** From Group Listing, click **Endocrine Tissue Slides**. To view slides without labels, click the **Labels Off** button at the bottom right of the monitor.

Thyroid Gland

The thyroid gland regulates metabolism by regulating the secretion of the hormones T_3 and T_4 (thyroxine) into the blood.

Click slide 1.
The gland is composed of fluid-filled **(colloid)** spheres, called **follicles,** formed by a simple epithelium that can be squamous to columnar depending upon the gland's activity. The colloid stored in the follicles is primarily composed of a glycoprotein (thyroglobulin) that is synthesized and secreted by the follicular cells. Under the influence of the pituitary gland, the follicular cells take up the colloid, convert it into T_3 and T_4, and secrete the T_3 and T_4 into an extensive capillary network. A second population of cells, parafollicular (C) cells (not shown), may be found scattered through the follicular epithelium, but often are present in the connective tissue between follicles. The pale-staining parafollicular cells secrete the protein hormone calcitonin.

Why is the thyroid gland considered to be an endocrine organ?

____

What hormone secreted by the pituitary gland controls the synthesis and secretion of T_3 and T_4 (thyroxine)?

____

What is the function of calcitonin?

____

Ovary

The ovary is an organ that serves both an exocrine function in producing eggs (ova) and an endocrine function in secreting the hormones estrogen and progesterone.

Click slide 2.
Grossly, the ovary is divided into a peripherally located **cortex** in which the oocytes (precursors to the ovulated egg) develop, and a central **medulla** in which connective tissue surrounds blood vessels, lymphatic vessels and nerves. The oocytes, together with supporting cells (granulosa cells), form the **ovarian follicles** seen in the cortex at various stages of development.

Click slide 3.
As an individual oocyte grows, **granulosa cells** proliferate from a single layer of cuboidal cells that surround the oocyte to a multicellular layer that defines a fluid-filled spherical follicle. In a mature follicle (Graafian follicle) the **granulosa cells** are displaced to the periphery of the fluid-filled **antrum,** except for a thin rim of granulosa cells **(corona radiata)** that encircles the **oocyte,** and a pedestal of granulosa cells (cumulus oophorus) that attaches the oocyte to the inner wall of the antrum.

Which cells of the ovarian follicle secrete estrogen?

____

Uterus

Click slides 4, 5, 6.
The uterus is a hollow muscular organ with three major layers: the **endometrium, myometrium,** and either an adventitia or a serosa.

The middle, myometrial layer of the uterine wall is composed of several layers of smooth muscle oriented in different planes.

Click slide 6.
The innermost (nearest the lumen) endometrial layer is further divided functionally into a superficial functional layer **(stratum functionalis)** and a deep basal layer **(stratum basalis).**

Click slide 4.
A simple columnar **epithelium** with both ciliated and nonciliated cells lines the surface of the **endometrium.** The endometrial connective tissue features an abundance of tubular endometrial **glands** that extend from the base to the surface of the layer. During the proliferative phase of the menstrual cycle, shown here, the endometrium becomes thicker as the glands and blood vessels proliferate.

Click slide 5.
In the secretory phase, the **endometrium** and its **glands** and blood vessels are fully expanded.

Click slide 6.
In the menstrual phase, the glands and blood vessels degenerate as the functional layer of the endometrium sloughs away. The deep basal layer **(stratum basalis)** is not sloughed and will regenerate the endometrium during the next proliferative phase.

Which layer of the endometrium is shed during the menstrual phase of the menstrual cycle?

____

What is the function of the deep basal layer (stratum basalis) of the endometrium?

____

What composes a serosa?

____

How does the serosa of the uterus, where present, differ from visceral peritoneum?

____

Pancreas

The pancreas is both an endocrine and an exocrine gland.

Click slide 7.
The exocrine portion is characterized by glandular **secretory units** (acini) formed by a simple epithelium of triangular or pyramidal cells that encircle a small central lumen. The central lumen is the direct connection to the duct system that conveys the exocrine secretions out of the gland. Scattered among the exocrine secretory units are the pale-staining clusters of cells that compose the endocrine portion of the gland. The cells that form these clusters, called **islets of Langerhans** cells (pancreatic islets), secrete a number of hormones, including insulin and glucagon.

Do the islets of Langerhans cells secrete their hormones into the same duct system used by the exocrine secretory cells?

____

Cardiovascular Tissue Review

From the PhysioEx main menu, select **Histology Review Supplement.** When the screen comes up, click **Select an Image Group.** From Group Listing, click **Cardiovascular Tissue Slides.** To view slides without labels, click the **Labels Off** button at the bottom right of the monitor.

Heart

The heart is a four-chambered muscular pump. Although its wall can be divided into three distinct histological layers (endocardium, myocardium, and epicardium), the cardiac muscle of the myocardium composes the bulk of the heart wall.

Click slide 1.
Contractile **cardiac muscle cells** (myocytes, myofibers) have the same striated appearance as skeletal muscle, but are branched rather than cylindrical in shape and have one (occasionally two) **nucleus** (myonucleus) rather than many. The cytoplasmic **striations** represent the same organization of myofilaments (sarcomeres) and alignment of sarcomeres as in skeletal muscle, and the mechanism of contraction is the same. The **intercalated disk,** however, is a feature unique to cardiac muscle. The densely stained structure is a complex of intercellular junctions (desmosome, gap junction, fascia adherens) that structurally and functionally link cardiac muscle cells end to end.

A second population of cells in the myocardium composes the noncontractile intrinsic conduction system (nodal system). Although cardiac muscle is autorhythmic, meaning it has the ability to contract involuntarily in the absence of extrinsic innervation provided by the nervous system, it is the intrinsic conduction system that prescribes the rate and orderly sequence of contraction. Extrinsic innervation only modulates the inherent activity.

Click slide 2.
Of the various components of the noncontractile intrinsic conduction system, **Purkinje fibers** are the most readily observed histologically. They are particularly abundant in the ventricular myocardium and are recognized by their very pale-staining cytoplasm and larger diameter.

The connective tissue component of cardiac muscle is relatively sparse and lacks the organization present in skeletal muscle.

Which component of the intercalated disk is a strong intercellular junction that functions to keep cells from being pulled apart during contraction?

____

What is a functional syncytium?

____

____

Which component of the intercalated disk is a junction that provides the intercellular communication required for the myocardium to perform as a functional syncytium?

____

Blood Vessels

Blood vessels form a system of conduits through which life-sustaining blood is conveyed from the heart to all parts of the body and back to the heart again.

Click slide 3.
Generally, the wall of every vessel is described as being composed of three layers or *tunics.* The **tunica intima,** or *tunica interna,* a simple squamous endothelium and a small amount of subjacent loose connective tissue, is the innermost layer adjacent to the vessel lumen. Smooth muscle and elastin are the predominant constituents of the middle **tunica media,** and the outermost **tunica adventitia,** or *tunica externa,* is a connective tissue layer of variable thickness that provides support and transmits smaller blood and lymphatic vessels and nerves. The thickness of each tunic varies widely with location and function of the vessel. **Arteries,** subjected to considerable pressure fluctuations, have thicker walls overall, with the tunica media being thicker than the tunica adventitia. **Veins,** in contrast, are subjected to much lower pressures and have thinner walls overall, with the tunica adventitia often outsizing the tunica media. Because thin-walled veins conduct blood back to the heart against gravity, valves (not present in arteries) also are present at intervals to prevent backflow. In capillaries, where exchange occurs between the blood and tissues, the tunica intima alone composes the vessel wall.

The tunica media of the aorta would have a much greater proportion of what type of tissue than a small artery?

____

In general, which vessel would have a larger lumen, an artery or its corresponding vein?

____

Why would the tunica media and tunica adventitia not be present in a capillary?

____

Respiratory Tissue Review

From the PhysioEx main menu, select **Histology Review Supplement.** When the screen comes up, click **Select an Image Group.** From Group Listing, click **Respiratory Tissue Slides.** To view slides without labels, click the **Labels Off** button at the bottom right of the monitor.

The respiratory system serves both to conduct oxygenated air deep into the lungs and to exchange oxygen and carbon dioxide between the air and the blood. The trachea, bronchi, and bronchioles are the part of the system of airways that conduct air into the lungs.

Click slide 2.
The trachea and bronchi are similar in morphology. Their lumens are lined by **pseudostratified columnar ciliated epithelium** with **goblet cells** (respiratory epithelium), underlain by a connective tissue **lamina propria** and a deeper connective tissue submucosa with coiled sero-mucous glands that open onto the surface lining of the airway lumen.

Click slide 1.
Deep to the submucosa are the **hyaline cartilage rings** that add structure to the wall of the airway and prevent its collapse. Peripheral to the cartilage is a connective tissue adventitia. The **sero-mucous glands** are also visible in this slide.

Click slide 3.
The bronchioles, in contrast, are much smaller in diameter with a continuous layer of **smooth muscle** in place of the cartilaginous reinforcements. A gradual decrease in the height of the epithelium to **simple columnar** also occurs as the bronchioles decrease in diameter. Distally the bronchioles give way to the respiratory bronchioles, alveolar ducts, alveolar sacs, and alveoli in which gas exchange occurs. In the respiratory bronchiole the epithelium becomes simple cuboidal and the continuous smooth muscle layer is interrupted at intervals by the presence of alveoli inserted into the bronchiolar wall.

Click slide 4.
Although some exchange occurs in the respiratory bronchiole, it is within the **alveoli** of the alveolar ducts and **sacs** that the preponderance of gas exchange transpires. Here the walls of the alveoli, devoid of smooth muscle, are reduced in thickness to the thinnest possible juxtaposition of simple squamous alveolar cell to simple squamous capillary endothelial cell.

What are the primary functions of the respiratory epithelium?

____

____

Why doesn't gas exchange occur in bronchi?

____

What is the primary functional unit of the lung?

____

The alveolar wall is very delicate and subject to collapse. Why is there no smooth muscle present in its wall for support?

____

____

What are the three basic components of the air-blood barrier?

____

____

Digestive Tissue Review

From the PhysioEx main menu, select **Histology Review Supplement.** When the screen comes up, click **Select an Image Group.** From Group Listing, click **Digestive Tissue Slides.** To view slides without labels, click the **Labels Off** button at the bottom right of the monitor.

Salivary Gland

The digestive process begins in the mouth with the physical breakdown of food by mastication. At the same time salivary gland secretions moisten the food and begin to hydrolyze carbohydrates. The saliva that enters the mouth is a mix of serous secretions and mucus (mucin) produced by the three major pairs of salivary glands.

Click slide 1.
The **secretory units** of the salivary tissue shown here are composed predominantly of clusters of pale-staining mucus-secreting cells. More darkly stained serous cells cluster to form a **demilune** (half moon) adjacent to the **lumen** and contribute a clear fluid to the salivary secretion. Salivary secretions flow to the mouth from the respective glands through a well-developed **duct** system.

Are salivary glands endocrine or exocrine glands?

____

Which salivary secretion, mucous or serous, is more thin and watery in consistency?

____

Esophagus

Through contractions of its muscular wall (peristalsis), the esophagus propels food from the mouth to the stomach. Four major layers are apparent when the wall of the esophagus is cut in transverse section:

Click slide 2.
1. The **mucosa** layer adjacent to the lumen consists of a nonkeratinized **stratified squamous epithelium,** its immediately subjacent connective tissue (lamina propria) containing blood vessels, nerves, lymphatic vessels, and cells of the immune system, and a thin smooth muscle layer (muscularis mucosa) that forms the boundary between the mucosa and the submucosa. Because this slide is a low magnification view, it is not possible to discern all parts of the mucosa, nor the boundary between it and the submucosa.

2. The **submucosa** is a layer of connective tissue of variable density, traversed by larger caliber vessels and nerves, that houses the mucus-secreting esophageal **glands** whose secretions protect the epithelium and further lubricate the passing food bolus.

3. Much of the substance of the esophageal wall consists of both circumferentially and longitudinally oriented layers of muscle called the **muscularis externa.** The muscularis externa is composed of skeletal muscle nearest the mouth, smooth muscle nearest the stomach, and a mix of both skeletal and smooth muscle in between.

4. The outermost layer of the esophagus is an adventitia for the portion of the esophagus in the thorax, and a serosa after the esophagus penetrates the diaphragm and enters the abdominal cavity.

Click slide 3.
Here we can see the abrupt change in epithelium at the gastroesophageal junction, where the **esophagus** becomes continuous with the **stomach.**

Briefly explain the difference between an adventitia and a serosa.

____

____

Stomach

The wall of the stomach has the same basic four-layered organization as that of the esophagus.

Click slide 4.
The **mucosa** of the stomach consists of a simple columnar epithelium, a thin connective tissue lamina propria, and a thin **muscularis mucosa.** The most significant feature of the stomach mucosa is that the epithelium invaginates deeply into the lamina propria to form superficial **gastric pits** and deeper **gastric glands.** Although the epithelium of the stomach is composed of a variety of cell types, each with a unique and important function, only three are mentioned here.

Click slide 5.
The **surface mucous cells** are simple columnar cells that line the **gastric pits** and secrete mucus continuously onto the surface of the epithelium. The large round pink- to red-stained **parietal cells** that secrete HCl line the upper half of the gastric glands, and more abundant in the lower half of the gastric glands are the chief cells (not shown), usually stained blue, that secrete pepsinogen (a precursor to pepsin).

Click slide 4 again.
The submucosa is similar to that of the esophagus, but without glands. The muscularis externa has the two typical circumferential and longitudinal layers of smooth muscle, plus an extra layer of smooth muscle oriented obliquely. The stomach's outermost layer is a serosa.

What is the function of the mucus secreted by surface mucous cells?

____

Small Intestine

The key to understanding the histology of the small intestine lies in knowing that its major function is absorption. To that end, its absorptive surface area has been amplified greatly in the following ways:

1. The mucosa and submucosa are thrown into permanent folds (plicae circulares).

2. Fingerlike extensions of the lamina propria form **villi** (singular: **villus**) that protrude into the intestinal lumen *(click slide 7).*

3. The individual **simple columnar epithelial** cells (enterocytes) that cover the villi have **microvilli** (a **brush border**), tiny projections of apical plasma membrane to increase their absorptive surface area *(click slide 6).*

Click slide 7.
Although all three segments of the small intestine (duodenum, jejunum, and ileum) possess villi and tubular **crypts** of Lieberkühn that project deep into the mucosa between villi, some unique features are present in particular segments. For example, large mucous glands **(Brunner's glands)** are present in the submucosa of the duodenum. In addition, permanent aggregates of lymphatic tissue **(Peyer's patches)** are a unique characteristic of the ileum *(click slide 8).*

Aside from these specific features and the fact that the height of the villi vary from quite tall in the duodenum to fairly short in the terminal ileum, the overall morphology of mucosa, submucosa, muscularis externa, and serosa is quite similar in all three segments.

Why is it important for the duodenum to add large quantities of mucus (from Brunner's glands) to the partially digested food entering it from the stomach?

____

Colon

Click slide 9.
The four-layered organization is maintained in the wall of the colon, but the colon has no villi, only **crypts** of Lieberkühn. Simple columnar epithelial cells (enterocytes with microvilli) are present to absorb water from the digested food mass, and the numbers of mucous **goblet cells** are increased substantially, especially toward the distal end of the colon.

Why is it important to have an abundance of mucous goblet cells in the colon?

____

Liver

The functional tissue of the liver is organized into hexagonally shaped cylindrical lobules, each delineated by connective tissue.

Click slide 11.
Within the lobule, large rounded **hepatocytes** form linear cords that radiate peripherally from the center of the lobule at the **central vein** to the surrounding connective tissue. Blood **sinusoids** lined by simple squamous endothelial cells and darkly stained **phagocytic Kupffer cells** are interposed between cords of hepatocytes in the same radiating pattern.

Click slide 10.
Located in the surrounding connective tissue, roughly at the points of the hexagon where three lobules meet, is the **portal triad** (portal canal).

Click slide 12.
The three constituents of the portal triad include a branch of the **hepatic artery,** a branch of the hepatic **portal vein,** and a **bile duct.** Both the hepatic artery and portal vein empty their oxygen-rich blood and nutrient-rich blood, respectively, into the sinusoids. This blood mixes in the sinusoids and flows centrally in between and around the hepatocytes toward the central vein. Bile, produced by hepatocytes, is secreted into very small channels (bile canaliculi) and flows peripherally (away from the central vein) to the bile duct. Thus, the flow of blood is peripheral to central in a hepatic lobule, while the bile flow is central to peripheral.

What general type of cell is the phagocytic Kupffer cell?

____

Blood in the portal vein flows directly from what organs?

____

What is the function of bile in the digestive process?

____

Pancreas

Click slide 13.
The exocrine portion of the pancreas synthesizes and secretes pancreatic enzymes. The individual exocrine **secretory unit,** or acinus, is formed by a group of pyramidal-shaped pancreatic acinar cells clustered around a central lumen into which they secrete their products. A system of pancreatic ducts then transports the enzymes to the duodenum where they are added to the lumen contents to further aid digestion. The groups of pale-staining cells are the endocrine **islets of Langerhans** (pancreatic islet) cells.

Renal Tissue Review

From the PhysioEx main menu, select **Histology Review Supplement.** When the screen comes up, click **Select an Image Group.** From Group Listing, click **Renal Tissue Slides.** To view slides without labels, click the **Labels Off** button at the bottom right of the monitor.

The many functions of the kidney include filtration, absorption, and secretion. The kidney filters the blood of metabolic wastes, water, and electrolytes, and reabsorbs most of the water and sodium ions filtered to regulate and maintain the body's fluid volume and electrolyte balance. The kidney also plays an endocrine role in secreting compounds that increase blood pressure and stimulate red blood cell production.

The uriniferous tubule is the functional unit of the kidney. It consists of two components: the nephron to filter and the collecting tubules and ducts to carry away the filtrate.

Click slide 1.
The nephron itself consists of the **renal corpuscle,** an intimate association of the glomerular capillaries **(glomerulus)** with the cup-shaped Bowman's capsule, and a single elongated renal tubule consisting of segments regionally and sequentially named the **proximal convoluted tubule (PCT),** the descending and ascending segments of the loop of Henle, and the distal convoluted tubule (DCT).

Click slide 2.
A closer look at the renal corpuscle shows both the simple squamous epithelium of the outer layer (parietal layer) of the **Bowman's capsule** (glomerular capsule), and the specialized inner layer (visceral layer) of **podocytes** that extend footlike processes to completely envelop the capillaries of the renal glomerulus. Processes of adjacent podocytes interdigitate with one another, leaving only small slits (filtration slits) between the processes through which fluid from the blood is filtered. The filtrate then flows into the **urinary space** that is directly continuous with the first segment of the renal tubule, the **PCT.** The PCT is lined by robust cuboidal cells equipped with microvilli to greatly increase the surface area of the side of the cell facing the lumen.

Click slide 3.
In the **loop of Henle,** lining cells are simple squamous to simple cuboidal. The DCT cells are also simple cuboidal but are usually much smaller than those of the PCT. The sparse distribution of microvilli, if present at all, on the cells of the DCT relates to their lesser role in absorption. The DCT is continuous directly with the **collecting tubules** and collecting ducts that drain the filtrate out of the kidney.

The large renal artery and its many subdivisions provide an abundant blood supply to the kidney. The smallest distal branches of the renal artery become the afferent arterioles that directly supply the capillaries of the glomerulus. In a unique situation, blood from the glomerular capillaries passes into the efferent arteriole rather than into a venule. The efferent arteriole then perfuses two more capillary beds, the peritubular capillary bed and vasa recta that provide nutrient blood to the kidney tissue itself, before ultimately draining into the renal venous system.

In which segment of the renal tubule does roughly 75–80% of reabsorption occur?

___

How are proximal convoluted tubule (PCT) cells similar to enterocytes of the small intestine?

___

What are the three layers through which the filtrate must pass starting from inside the glomerular capillary through to the urinary space?

___

Under normal circumstances in a healthy individual, would red blood cells or any other cells be present in the renal filtrate?

___

In addition to providing nutrients to the kidney tubules, what is one other function of the capillaries in the peritubular capillary bed?

___